DNA Repair Protocols

METHODS IN MOLECULAR BIOLOGY™

John M. Walker, Series Editor

161. **Cytoskeleton Methods and Protocols**, edited by *Ray H. Gavon, 2001*
160. **Nuclease Methods and Protocols,** edited by *Catherine H. Schein, 2000*
159. **Amino Acid Analysis Protocols,** edited by *Catherine Cooper, Nicole Packer, and Keith Williams, 2000*
158. **Gene Knockoout Protocols,** edited by *Martin J. Tymms and Ismail Kola, 2000*
157. **Mycotoxin Protocols,** edited by *Mary W. Trucksess and Albert E. Pohland, 2000*
156. **Antigen Processing and Presentation Protocols,** edited by *Joyce C. Solheim, 2000*
155. **Adipose Tissue Protocols,** edited by *Gérard Ailhaud, 2000*
154. **Connexin Methods and Protocols,** edited by *Roberto Bruzzone and Christian Giaume, 2000*
153. **Neuropeptide Y Protocols**, edited by *Ambikaipakan Balasubramaniam, 2000*
152. **DNA Repair Protocols:** *Prokaryotic Systems*, edited by *Patrick Vaughan, 2000*
151. **Matrix Metalloproteinase Protocols,** edited by *Ian M. Clark, 2000*
150. **Complement Methods and Protocols,** edited by *B. Paul Morgan, 2000*
149. **The ELISA Guidebook**, edited by *John R. Crowther, 2000*
148. **DNA–Protein Interactions:** *Principles and Protocols* **(2nd ed.)**, edited by *Tom Moss, 2000*
147. **Affinity Chromatography:** *Methods and Protocols*, edited by *Pascal Bailon, George K. Ehrlich, Wen-Jian Fung, and Wolfgang Berthold, 2000*
146. **Protein and Peptide Analysis:** *New Mass Spectrometric Applications*, edited by *John R. Chapman, 2000*
145. **Bacterial Toxins: *Methods and Protocols***, edited by *Otto Holst, 2000*
144. **Calpain Methods and Protocols**, edited by *John S. Elce, 2000*
143. **Protein Structure Prediction:** *Methods and Protocols*, edited by *David Webster, 2000*
142. **Transforming Growth Factor-Beta Protocols**, edited by *Philip H. Howe, 2000*
141. **Plant Hormone Protocols**, edited by *Gregory A. Tucker and Jeremy A. Roberts, 2000*
140. **Chaperonin Protocols**, edited by *Christine Schneider, 2000*
139. **Extracellular Matrix Protocols**, edited by *Charles Streuli and Michael Grant, 2000*
138. **Chemokine Protocols**, edited by *Amanda E. I. Proudfoot, Timothy N. C. Wells, and Christine Power, 2000*
137. **Developmental Biology Protocols, Volume III**, edited by *Rocky S. Tuan and Cecilia W. Lo, 2000*
136. **Developmental Biology Protocols, Volume II,** edited by *Rocky S. Tuan and Cecilia W. Lo, 2000*
135. **Developmental Biology Protocols, Volume I,** edited by *Rocky S. Tuan and Cecilia W. Lo, 2000*
134. **T Cell Protocols:** *Development and Activation*, edited by *Kelly P. Kearse, 2000*
133. **Gene Targeting Protocols,** edited by *Eric B. Kmiec, 2000*
132. **Bioinformatics Methods and Protocols,** edited by *Stephen Misener and Stephen A. Krawetz, 2000*
131. **Flavoprotein Protocols,** edited by *S. K. Chapman and G. A. Reid, 1999*
130. **Transcription Factor Protocols,** edited by *Martin J. Tymms, 2000*
129. **Integrin Protocols,** edited by *Anthony Howlett, 1999*
128. **NMDA Protocols,** edited by *Min Li, 1999*
127. **Molecular Methods in Developmental Biology:** Xenopus *and Zebrafish*, edited by *Matthew Guille, 1999*
126. **Adrenergic Receptor Protocols,** edited by *Curtis A. Machida, 2000*
125. **Glycoprotein Methods and Protocols:** *The Mucins*, edited by *Anthony P. Corfield, 2000*
124. **Protein Kinase Protocols,** edited by *Alastair D. Reith, 2000*
123. ***In Situ* Hybridization Protocols (2nd ed.),** edited by *Ian A. Darby, 2000*
122. **Confocal Microscopy Methods and Protocols,** edited by *Stephen W. Paddock, 1999*
121. **Natural Killer Cell Protocols:** *Cellular and Molecular Methods*, edited by *Kerry S. Campbell and Marco Colonna, 2000*
120. **Eicosanoid Protocols,** edited by *Elias A. Lianos, 1999*
119. **Chromatin Protocols,** edited by *Peter B. Becker, 1999*
118. **RNA–Protein Interaction Protocols,** edited by *Susan R. Haynes, 1999*
117. **Electron Microscopy Methods and Protocols,** edited by *M. A. Nasser Hajibagheri, 1999*
116. **Protein Lipidation Protocols,** edited by *Michael H. Gelb, 1999*
115. **Immunocytochemical Methods and Protocols (2nd ed.),** edited by *Lorette C. Javois, 1999*
114. **Calcium Signaling Protocols,** edited by *David G. Lambert, 1999*
113. **DNA Repair Protocols:** *Eukaryotic Systems*, edited by *Daryl S. Henderson, 1999*
112. **2-D Proteome Analysis Protocols,** edited by *Andrew J. Link, 1999*
111. **Plant Cell Culture Protocols,** edited by *Robert D. Hall, 1999*
110. **Lipoprotein Protocols,** edited by *Jose M. Ordovas, 1998*
109. **Lipase and Phospholipase Protocols,** edited by *Mark H. Doolittle and Karen Reue, 1999*
108. **Free Radical and Antioxidant Protocols,** edited by *Donald Armstrong, 1998*
107. **Cytochrome P450 Protocols,** edited by *Ian R. Phillips and Elizabeth A. Shephard, 1998*
106. **Receptor Binding Techniques,** edited by *Mary Keen, 1999*
105. **Phospholipid Signaling Protocols,** edited by *Ian M. Bird, 1998*
104. **Mycoplasma Protocols,** edited by *Roger J. Miles and Robin A. J. Nicholas, 1998*
103. ***Pichia* Protocols,** edited by *David R. Higgins and James M. Cregg, 1998*
102. **Bioluminescence Methods and Protocols,** edited by *Robert A. LaRossa, 1998*
101. **Mycobacteria Protocols,** edited by *Tanya Parish and Neil G. Stoker, 1998*
100. **Nitric Oxide Protocols**, edited by *Michael A. Titheradge, 1998*
99. **Stress Response:** *Methods and Protocols*, edited by *Stephen M. Keyse, 2000*
98. **Forensic DNA Profiling Protocols**, edited by *Patrick J. Lincoln and James M. Thomson, 1998*
97. **Molecular Embryology:** *Methods and Protocols*, edited by *Paul T. Sharpe and Ivor Mason, 1999*

METHODS IN MOLECULAR BIOLOGY™

DNA Repair Protocols

Prokaryotic Systems

Edited by

Pat Vaughan

HiberGen Ltd. and University College, Cork, Ireland

Humana Press **Totowa, New Jersey**

Softcover reprint of the hardcover 1st edition 2000
999 Riverview Drive, Suite 208
Totowa, New Jersey 07512

This publication is printed on acid-free paper. ∞
ANSI Z39.48-1984 (American Standards Institute)
Permanence of Paper for Printed Library Materials.

Cover design by Patricia F. Cleary.

For additional copies, pricing for bulk purchases, and/or information about other Humana titles, contact Humana at the above address or at any of the following numbers: Tel.: 973-256-1699; Fax: 973-256-8341; E-mail: humana@humanapr.com; or visit our Website: http://humanapress.com

ISBN 978-1-61737-112-7 e-ISBN 978-1-59259-068-1

10 9 8 7 6 5 4 3 2 1

Library of Congress Cataloging in Publication Data

Main entry under title:

Methods in molecular biology™.

DNA repair protocols : prokaryotic systems / edited by Patrick Vaughan.
p. cm.—(Methods in molecular biology; 152)
Includes bibliographical references and index.

1. DNA repair—Laboratory manuals. I. Vaughan, Patrick. II. Methods in molecular biology (Totowa, NJ); v. 152.

QH467.D63 2000

572.8'6459293—dc21

99-086507

Preface

When setting out to decide on the content of *DNA Repair Protocols: Prokaryotic Systems*, I was conscious of the need to portray the vast array of pathways and enzymatic activities that are part of the discipline of DNA repair. In addition to the classical DNA repair activities, I wanted to convey the significant interest that has been generated in recent years in the use of the proteins and repair systems as research tools, much like the use of restriction enzymes over the last few decades. Therefore, in addition to chapters detailing protocols for investigating specific repair activities, I have included several chapters in this book on the applied use of DNA repair proteins and systems.

The many years of research on bacterial DNA repair systems have allowed us to really understand the majority of DNA repair pathways in bacterial cells. Building on this knowledge, research has lead to major advances in understanding mammalian DNA repair and uncovered its links to human disease, such as DNA mismatch repair and colon cancer, nucleotide excision repair and xeroderma pigmentosum, DNA helicase function in Bloom's syndrome, and so on. Such have been the advances that *Science* magazine identified the collective DNA repair systems as its "Molecule of the Year" in 1994. Because interest in DNA repair continues today, our hope is that new researchers in the field will find this book useful to bring themselves up to date on the basis of cellular DNA repair activities, be their interest in prokaryotic or eukaryotic systems. Also included here is the investigation of DNA repair in thermophilic bacteria. It is interesting to note that in these cells, repair protein needs to sustain activity at high temperatures in an environment where DNA is inherently more susceptible to damage. Therefore it is clear that more researchers will be attracted to the investigation of the classical DNA repair activities in these organisms, both because of the novel nature of the molecular environment and of the possible biotechnological use of these thermostable enzymes.

As stated earlier, I have included many protocols detailing the study of various repair activities found in cells, and contrasted this with a number of chapters where researchers have taken these same DNA repair enzymes and systems and used them as tools or reagents in various techniques. Several chapters deal with the use of repair proteins in mutation detection. Mutation and polymorphism detection is a growing field of research. It is central to the frantic race to uncover the multitude of genetic variations impacting on human and animal health, both in the search for disease genes, drug response and drug metabolism genes, and to breeding and trait selection in animals and plants. The use of DNA repair proteins as biotechnological tools is invaluable in this effort owing to the specificity of these proteins and our in-depth understanding of their action.

I thank all the authors of this volume for their diligence and cooperation in putting together this book and wish them continued success in their research. I hope you the reader find our collection interesting, informative, and practical.

Pat Vaughan

Contents

Contributors

Jeffrey J. Babon • *Mutation Research Centre, St. Vincent's Hospital, Melbourne, Victoria, Australia*

Peter Brooks • *Génoscope, Centre National de Séquençage, Evry Cédex, France*

Richard G. H. Cotton • *Mutation Research Centre, St. Vincent's Hospital, Melbourne, Victoria, Australia*

Claire G. Cupples • *Department of Biology, Concordia University, Montréal, Québec, Canada*

Alan Dean • *Gene Check Inc., Fort Collins, CO*

Lance J. Ferrin • *Division of Gastroenterology, Department of Medicine, University of Minnesota, Minneapolis, MN*

William A. Franklin • *Department of Radiology and Radiation Oncology, Albert Einstein College of Medicine, Bronx, NY*

Susana Gonzalez • *Department of Microbiology, Mount Sinai School of Medicine, New York, NY*

Frank G. Harmon • *Division of Biological Sciences, Sections of Microbiology and Molecular and Cellular Biology, University of California, Davis, CA*

Douglas A. Julin • *Department of Chemistry and Biochemistry, University of Maryland, College Park, MD*

Stephen C. Kowalczykowski • *Section of Microbiology, Division of Biological Sciences, University of California, Davis, CA*

A-Lien Lu • *Department of Biochemistry and Molecular Biology, School of Medicine, University of Maryland, Baltimore, MD*

Georgina Macintyre • *Department of Biology, Concordia University, Montréal, Québec, Canada*

Matthew McKenzie • *Mutation Research Centre, St. Vincent's Hospital, Melbourne, Victoria, Australia*

Geoffrey P. Margison • *CRC Department of Carcinogenesis, Paterson Institute for Cancer Research, Christie Hospital (NHS) Trust, Manchester, UK*

IVAN MATIC • *INSERM E9916, Faculte de Medicine–Necker–"Enfants malades," Universite Rene Descartes–Paris V, Paris Cédex, France*
GERALDINE M. O'GRADY • *Department of Biochemistry, University College, Cork, Ireland*
MIROSLAV RADMAN • *INSERM E9916, Faculte de Medicine–Necker–"Enfants malades," Universite Rene Descartes–Paris V, Paris Cédex, France*
MARGARITA SANDIGURSKY • *Department of Radiology and Radiation Oncology, Albert Einstein College of Medicine, Bronx, NY*
FRANÇOIS TADDEI • *INSERM E9916, Faculte de Medicine–Necker–"Enfants malades," Universite Rene Descartes–Paris V, Paris Cédex, France*
PAT VAUGHAN • *HiberGen, Ltd., and Department of Biochemistry and NFBC, University College, Cork, Ireland*
ROBERT WAGNER • *Gene Check Inc., Fort Collins, CO*
AMANDA J. WATSON • *CRC Department of Carcinogenesis, Paterson Institute for Cancer Research, Christie Hospital (NHS) Trust, Manchester, UK*
JAMES G. WETMUR • *Department of Microbiology, Mount Sinai School of Medicine, New York, NY*

Part I

Assays to Detect and Quantitate DNA Repair Activity

1

Repair of A/G and A/8-oxoG Mismatches by MutY Adenine DNA Glycosylase

A-Lien Lu

1. Introduction

Cellular and organism aging have been correlated with accumulated DNA damage *(1,2)*. 8-oxo-7,8-dihydrodeoxyguanine (8-oxoG or GO) is one of the most stable products of oxidative DNA damage. The formation of GO in DNA, if not repaired, can lead to misincorporation of A opposite to the GO lesion and result in G:C to T:A transversions *(3–6)*. In *Escherichia coli*, a family of enzymes, MutY, MutM, and MutT, is involved in defending against the mutagenic effects of GO lesions *(7–9)*. The *E. coli* MutY is an adenine glycosylase active on DNA containing A/GO, A/G, and A/C mismatches *(7,10–15)* and also has a weak guanine glycosylase activity on G/GO-containing DNA *(15a,15b)*. MutY removes misincorporated adenines paired with GO lesions and reduces the GO mutational effects. The 39-kDa MutY protein from *E. coli* is an iron-sulfur protein. The MutY protein was shown by Tsai-Wu et al. *(16)* to have both DNA glycosylase and apurinic/apyrimidinic (AP) lyase activities. Recent results show that MutY and the N-terminal catalytic domain can be trapped in a stable covalent enzyme-DNA intermediate in the presence of sodium borohydride *(17–19)* and support that MutY contains both DNA glycosylase and AP lyase activities. The DNA glycosylase activity removes the adenine bases from the A/GO, A/G, and A/C mismatches *(16)* and the AP lyase activity cleaves the first phosphodiester bond 3' to the AP site *(12,16)*. Apparent dissociation constants are 0.066, 5.3, and 15 n*M* for A/GO-, A/G-, and A/C-containing DNA, respectively *(20)*.

<u>M</u>ut<u>Y</u> <u>h</u>omologous (MYH) activities have been identified in human HeLa *(21)*, calf thymus *(22)*, and fission yeast *Schizosaccharomyces pombe (23)*.

From: *Methods in Molecular Biology, vol. 152: DNA Repair Protocols: Prokaryotic Systems*
Edited by: P. Vaughan © Humana Press Inc., Totowa, NJ

The recombinant human MYH from the cloned cDNA has been expressed and partially characterized ***(24a–c)***. A human cDNA of putative hMYH has been cloned ***(24)***. These MYH proteins share high-sequence homology and similar mechanisms with the *E. coli* MutY protein ***(21,24)***. The high homology of MutY homologs among different organisms suggests important roles in their cellular functions.

Genetic mutations can be detected by MutY protein ***(25,26)*** based on its specific binding and nicking of DNA heteroduplexes containing an A/G or A/C mismatch. In this mismatch repair enzyme cleavage (MREC) method, DNA fragments amplified from normal and mutated genes by polymerase chain (PCR) are mixed and annealed to create base/base mismatches for cleavage by repair enzymes. MutY can detect A:T–C:G transversions and G:C–A:T transitions. The method is powerful and sensitive.

2. Materials

2.1. Reagents and Buffers

1. 10X MutY reaction buffer: 200 m*M* Tris-HCl, pH 7.6, 800 m*M* NaCl, 10 m*M* dithiothreitol (DTT), 10 m*M* ethylendiaminetetraacetic acid (EDTA), 29% (v/w) glycerol.
2. 10X MYH reaction buffer: 100 m*M* Tris-HCl, pH 7.6, 5 m*M* DTT, 5 m*M* EDTA, 15% (v/w) glycerol.
3. MutY storage/dilution buffer: 20 m*M* potassium phosphate, pH 7.4, 1.5 m*M* DTT, 0.1 m*M* EDTA, 50 m*M* KCl, 200 µg/mL bovine serum albumin (BSA), and 50% glycerol.
4. 10X hybridization buffer: 70 m*M* Tris-HCl, pH 7.6, 70 m*M* $MgCl_2$, and 500 m*M* NaCl.
5. 5X Klenow buffer: 250 m*M* Tris-HCl, pH 7.6, 25 m*M* $MgCl_2$, 25 m*M* β-mercaptoethanol, 0.1 m*M* dGTP, and 0.1 m*M* dTTP.
6. 10X kinase buffer: 500 m*M* Tris-HCl, pH 7.6, 100 m*M* $MgCl_2$, 50 m*M* DTT, 1 m*M* spermidine, and 1 m*M* EDTA.
7. 10X DNA dye: 60% glycerol, 50 m*M* EDTA, 0.5% sodium dodecyl sulfate (SDS), 0.05% xylene cyanol, and 0.05% bromophenol blue.
8. Sequencing dye: 90% formamide, 10 m*M* EDTA, 0.1% xylene cyanol, and 0.1% bromophenol blue.
9. 5X SDS-polyacrylamide gel electophoresis (PAGE) dye: 155 m*M* Tris-HCl, pH 6.8, 25% (v/v) glycerol, 5% (w/v) SDS, 0.5 mg/mL bromophenol blue, and 5% (v/v) β-mercaptoethanol.
10. Klenow fragment of DNA polymerase I (New England BioLabs).
11. Polynucleotide kinase (New England BioLabs).
12. Poly(dI-dC): 200 µL at 10 µg/mL (Parmacia Biotech).
13. TBE buffer: 50 m*M* Tris-borate, pH 8.3, and 1 m*M* EDTA.
14. SDS-PAGE running buffer: 25 m*M* Tris-base, 192 m*M* glycine, and 1% SDS.

15. $TE_{0.1}$ buffer: 10 m*M* Tris-HCl, pH 7.6, 0.1 m*M* EDTA.
16. Quick-spin column (Boehringer Mannheim).
17. [α-P^{32}] dCTP and [γ-P^{32}] ATP at 3000 Ci/mmol from NEN.
18. Diethylaminoethyl (DEAE)-81 paper (Whatman, cut into 1.2 × 1.2-cm squares).
19. GF/C filter (Whatman, 2.4-cm circle).
20. Coomassie stain: 0.25% (w/v) Coomassie brillant blue R250 in 50% methanol and 10% acetic acid.
21. Buffer T: 50 m*M* Tris-HCl, pH 7.6, 0.1 m*M* EDTA, 0.5 m*M* DTT, and 0.1 m*M* phenylmethylsulfonyl fluoride (PMSF).
22. Buffer A: 20 m*M* potassium phosphate, pH 7.4, 0.5 m*M* DTT, 0.1 m*M* EDTA, and 0.1 m*M* PMSF.
23. Buffer B: 0.01 *M* potassium phosphate, pH 7.4, 10 m*M* KCl, 0.5 m*M* DTT, 0.1 m*M* EDTA, and 0.1 m*M* PMSF.

2.2. DNA Substrates

2.2.1. Synthesis and Purification

The 19-mer oligonucleotides (*see sequences* in **Subheading 2.2.2.**) were synthesized at 0.2-µmol scale on an Applied Biosystems 381A automated synthesizer by using standard procedures. Phosphoramidite of 8-oxo-dG was purchased from Glen Research.

1. Load deprotected oligonucleotides (1 OD per 1 cm × 0.15-cm well) on a 14% sequencing gel ***(27)*** and run the gel at 600 V for 40 min.
2. Put the gel over a Whatman TLC plate (cat. no. 4410222) and shine it with short-wave UV from a hand UV lamp (UV shadowing). Excise the full-length bands (up to 10 OD) in a 15-mL Falcon centrifuge tube.
3. Crush the gel with a clean glass rod and add 10 mL of 1 *M* triethylammonium bicarbonate (TEAB), pH 7.0 to the tube, which is rotated overnight at 37°C.
4. Spin with a table-top centrifuge for 10 min and transfer the supernatant to a new tube.
5. Wash a C18 Sep-Pak column (Waters) with 10 mL each of 100% ethanol, 50% ethanol/50% 25 m*M* TEAB, and then 25 m*M* TEAB.
6. Load the eluted DNA onto the C18 Sep-Pak column.
7. Wash the column with 10 mL 25 m*M* TEAB and elute DNA with 2 mL of 40% ethanol/ 60% 25 m*M* TEAB.
8. Lyophilize the sample to dry and dissolve DNA with 1 mL of distilled water. Determine its concentration by A_{260} quantitation (1 OD = 33 µg/mL).

2.2.2. Annealing

1. Mix two complementary oligonucleotides in hybridization buffer in a 1.5-mL microtube (150 pmol each 15 µL of 10X hybridization buffer and water to 150 µL).
2. Heat at 90°C for 2 min and then the tube is placed on the top of a 25-mL beaker with 90°C water and cooled gradually to room temperature over more than 30 min. Heteroduplexes are constructed as follows.

```
5'-CCGAGGAATTAGCCTTCTG-3'
 3'-GCTCCTTAAGCGGAAGACG-5'

5'-CCGAGGAATTAGCCTTCTG-3'
 3'-GCTCCTTAAOCGGAAGACG-5'

5'-CCGAGGAATTAGCCTTCTG-3'
 3'-GCTCCTTAACCGGAAGACG-5'

5'-CCGAGGAATTCGCCTTCTG-3'
 3'-GCTCCTTAAGCGGAAGACG-5'
```

2.4. Apparatus

1. Sequencing gel apparatus (IBI STS 45i DNA sequencing unit cat. no. IB80000 or BRL cat. no. 21070-016 for 0.8-mm-thick spacer).
2. SDS-PAGE apparatus (Novex cat. no. EI9001).
3. Gel-shifting apparatus (BRL cat. no. 21070-024 for 1.5-mm-thick spacer).
4. Power supplies.
5. Desiccator.
6. Gel dryer.
7. X-ray film cassettes.
8. Microcentrifuge.
9. Water bath.
10. Beckman 70.1 Ti rotor and centrifuge.
11. Waters or Pharmacia FPLC system.
12. Table-top IEC clinical centrifuge.

3. Methods

3.1. Preparation of Labeled DNA Substrates

3.1.1. 3'-End Labeling Reaction

Oligonucleotides with A/G, A/GO, or A/C mismatches are substrates for MutY. Homoduplex with C:G is not a substrate and should also be used as a negative control for MutY nicking and binding.

1. To a microcentrifuge tube, add the following in order:

Sterile dH_2O	5.5 µL
5X Klenow buffer	3 µL
Duplex Oligonucleotide (1 pmol/mL)	1 µL
[α-^{32}P] dCTP at 3,000 Ci/mmol	5 µL
Klenow fragment (5 U/µL)	0.5 µL
Total	15 µL

2. Incubate the reaction for 30 min at room temperature. (At the same time, prepare G-25 column, *see below*.)
3. Then, add 1 µL of 0.5 *M* EDTA and 34 µL of $TE_{0.1}$ to stop the reaction.

4. Spot 0.5 µL onto a piece of square DEAE paper and then wash as described in **Subheading 3.1.4.**
5. Pass the rest of the sample through a Quick-Spin G-25 column as described in **steps 8-13** in **Subheading 3.1.3.**

3.1.2. 5'-End-Labeling Reaction

1. To a microcentrifuge tube, add the following in order:

Sterile dH_2O	6.8 µL
10X kinase buffer	2 µL
Oligonucleotide (single-stranded) (1 pmol/µL)	1 µL
[γ-^{32}P] ATP at 3,000 Ci/mmol	10 µL
T4 polynucleotide kinase (10 U/µL)	0.5 µL
Total	15 µL

2. Incubate the reaction for 30 min at 37°C.
3. Stop the reaction by heating at 65°C for 5 min.
4. Add 30 µL of $TE_{0.1}$.
5. Spot 0.5 µL onto a piece of square DEAE paper and then wash as described in **Subheading 3.1.4.**
6. Add 2 µL of 10X hybridization buffer and 2 pmol of the complementary strand of oligonucleotide.
7. Heat 90°C for 2 min and then cool gradually to room temperature over 30 min to form heteroduplexes.
8. Add 0.2 µL 10 m*M* each of the four dNTP and 0.5 µL of Klenow fragment. Incubate for 30 min.
9. Load the sample onto a Quick-Spin column (*see* **steps 8–13** in **Subheading 3.1.3.**)

3.1.3. Removal of Free Nucleotides

These procedures are modified from the manufacturers man ual.

1. Invert the Quick-Spin G25 column several times.
2. Remove the top and bottom caps.
3. Put one receiving tube in a 15-mL plastic tube and then the column.
4. Spin the assembly in a table-top IEC clinical centrifuge with swing buckets for 2 min.
5. Discard the solution.
6. Add 0.4 mL $TE_{0.1}$ on the top of the column, repeat **steps 4** and **5.**
7. Spin again for 2 min without adding buffer. Discard the solution and replace a new receiving tube.
8. Load 49.5 µL of labelled sample from **Subheadings 3.1.1.** or **3.1.2.** (Remember to spot 0.5 µL of sample onto DEAE paper to check incorporation, *see* **Subheading 3.1.4.**)
9. Spin for 4 min.
10. Carefully transfer the solution passed through the column into another tube and measure its volume.
11. Spot 0.5 µL onto a piece of square DEAE paper and follow the washing steps (*see* **Subheading 3.1.4.**) and spot 0.5 µL onto a GF/C filter paper, dry under a heat lamp, and count.

12. Store the rest in –20°C and make proper dilution to 1.8 fmol/μL according to **Subheading 3.1.5.**

3.1.4. Check Incorporation

1. Spot 0.5 μL of labeled DNA samples before and after the G-25 column onto pieces of DEAE paper. (Mark paper squares with a pencil.)
2. Wash DEAE papers with 200 mL of 0.25 *M* ammonium bicarbonate contained within a 1L beaker.
3. Shake 5 min with speed sufficient for the papers to float.
4. Carefully discard the washing solution into a radioactive waste jar.
5. Repeat **steps 2–4** two more times.
6. Wash DEAE papers with 200 mL of 95% ethanol, similar to **steps 2–4** three times.
7. Put papers on a sheet of aluminum foil and dry 10 min under a heat lamp.
8. Add 5 mL of scintillation cocktail and count.

*3.1.5. Determination of Specific Activity (*see **Note 1***)*

Pre-G-25 DEAE paper	A cpm/0.5 μL
Post-G-25 DEAE paper	B cpm/0.5 μL
Post-G-25 GF/C paper	C cpm/0.5 μL
Total cpm = cpm/pmol (pre-column)	T1 = A × 100
cpm/1.8 fmol	S = T1 × 0.0018
Total cpm (postcolumn)	T2 = 2B ↔ vol (post-column)
Recovery	T2/T1
% of cpm in DNA post-G25	B/C
Dilution to 1.8 fmol/μL	2B/S

3.2. MutY Enzyme Purification

1. Grow 12 L (four 3 L media in 6 L flasks) of *E. coli* JM109 cells harboring overproduction plasmid pJTW10-12 *(16)* to A_{590} of 0.7 in LB broth containing 50 mg/mL of ampicillin at 37°C.
2. Induce MutY production by adding IPTG to 0.4 m*M* and the cultures are continuously shaken overnight at 28°C (*see* **Note 2**).
3. Harvest cells by centrifugation in a GS3 rotor at 11,000*g* for 15 min. At the end, remove as much media as possible and scrape the cell paste in a 50-mL plastic tube that is stored at –80°C.
4. Before the enzyme purification, prepare all the required buffers (filtered through 45 μ membrane and autoclaved) and pack the columns at 4°C. All column chromatography is conducted in a Waters 650 FPLC system at 4°C and centrifugation is done at 65000*g* for 30 min (*see* **Note 3**).
5. Cells (40 g of cell paste) are resuspended in 120 mL of buffer T and disrupted with a bead beater (Biospec Products, Bartlesville, OK) using 0.1-mm glass beads (10 times 20 s blending and 10 s pulse).
6. Remove cell debris by centrifugation and carefully pour the supernatant to a graduated cylinder. The supernatant is then treated with 5% streptomycin sulfate in buffer T and stirred for 30 min (nucleic acids are precipitated out).

7. After centrifugation, the supernatant is collected as fraction I (235 mL).
8. Add ammonium sulfate (162 g) to fraction I. After stirring for 30 min, the protein is precipitated overnight.
9. After centrifugation, resuspend the protein pellet in 12 mL of buffer T and the sample is dialyzed against two changes of 1 L of buffer T for 3 h each.
10. The dialyzed protein sample is diluted four-fold with buffer A containing 50 m*M* KCl as fraction II (100 mL). Fraction II is loaded at flow rate of 2 mL per min onto a 30 mL phosphocellulose (Whatman P-11) column that has been equilibrated with buffer A containing 0.05 *M* KCl.
11. After washing with 60 mL of equilibration buffer, proteins are eluted with a 300-mL linear gradient of KCl (0.05–0.5 *M*) in buffer A. Fractions containing the A/G-specific nicking activity (*see* **Subheading 3.5.1.**) are pooled (fraction III, 67 mL) (those fractions should have brown color because MutY contains a Fe-S cluster).
12. Load fraction III onto a 20-mL hydroxylapatite column equilibrated with buffer B. After washing with 40 mL of equilibration buffer, the flowthrough and early elution fractions are pooled and dialyzed against buffer A containing 0.05 *M* KCl and 10% (vol/vol) glycerol for 2 h (fraction IV, 63 mL).
13. Fraction IV is loaded onto a 5-mL heparin-agarose column equilibrated with buffer A containing 0.05 *M* KCl and 10% glycerol. After washing with 10 mL of equilibration buffer, the column is developed with a 50-mL linear gradient of KCl (0.1–0.6 *M*) in buffer A with 10% glycerol. Fractions containing the MutY nicking activity, which eluted at 0.3 *M* KCl, are pooled (fraction V, 17 mL), are then divided into small aliquots and stored at –80°C. Protein concentration is determined by the Bradford method ***(28)***.

3.3. Preparation of Crude Cell Extracts

If pure MutY is not required for the experimental purpose, small-scale crude extracts can be obtained to check MutY repair activity ***(11)*** by the following procedures.

1. Grow 1 L of *E. coli* JM109 cells harboring overproduction plasmid pJTW10-12 ***(16)*** to A_{590} of 0.7 in LB broth containing 50 mg/mL of ampicillin at 37°C. Add IPTG to 0.4 m*M* to the culture and leave overnight at 28°C.
2. Harvest by centrifugation in a GSA rotor at 10,500*g* for 15 min. Remove media as much as possible.
3. The cells are resuspended in 2 mL of 0.05 *M* Tris-HCl, pH 7.6, 10% sucrose, transferred to a centrifuge tube for Beckman 70.1 Ti rotor, quickly frozen in a dry ice/ethanol bath, and stored at –80°C.
4. Cell suspensions are supplemented with 1.2 m*M* DTT, 0.15 *M* KCl, 0.23 mg/mL of lysozyme, kept on ice for 1 h, and heated at 37°C for a time sufficient to yield a final suspension temperature of 20°C.
5. Centrifuge at 100,000*g* in a Beckman 70.1 Ti rotor for 1 h at 4°C. Save the supernatant in a 15-mL Corex glass tube with a very small stirring bar.
6. Add solid ammonium sulfate (0.42 g/mL) to the supernatant. Stir for 20 min.

7. Collect the precipitate by centrifugation at 19,000*g* in SS34 rotor for 25 min.
8. Resuspend the pellet in 0.3 mL of 25 m*M* HEPES, pH 7.6, 0.1 m*m* EDTA, 2 m*M* DTT, 0.1 *M* KCl and dialyze the sample against the same buffer (2 × 250 mL) for 90 min. Check the conductivity of the sample by diluting 10 µL into 4 mL distilled water. The conductivity should be about 80 µS.
9. The protein sample is quickly frozen in small aliquots and stored at –80°C.

3.4. MutY Binding Assay

3.4.1. Binding Assay

The binding of MutY to DNA substrates is assayed by gel retardation (*see* **Notes 4** and **5**).

1. Prerun an 8% polyacrylamide nondenaturing gel in 1X TBE buffer at 150 V for more than 30 min.
2. To each reaction, add the following in order to a microcentrifuge tube.

Sterile dH_2O	14 µL
10X MutY reaction buffer	2 µL
10 µg/mL poly (dI-dC)	2 µL
3' end-labeled DNA (1.8 fmol)	1 µL
MutY (72 nM)	1 µL
Total	20 µL

   Assay both A/G- and C/G-containing DNA. A control incubation consisting of DNA only (no MutY protein) should also be run. Dilute MutY enzyme with storage/dilution buffer. Incubate all reactions at 37°C for 30 min.
3. Remove the reaction tubes from water bath and add 1.5 µL of 50% glycerol.
4. Load the entire reaction products onto the gel. Do not delay in loading the samples onto the gel. Also load into an adjacent well with 1X DNA dye in $TE_{0.1}$.
5. Run the gel at 10 V/cm until bromophenol blue has migrated more than half-way down the gel.
6. Remove the glass plates and transfer the gel onto 3MM filter paper.
7. Dry the gel in a gel dryer for 45 min and autoradiograph until the proper exposure is achieved. It takes 16 h for 3000 cpm of DNA. The free DNA migrates below the bromophenol blue and the MutY-bound complex migrates at a position near xylene cyanol.

3.4.2. K_d Determination

The apparent dissociation constants (K_d) of MutY and DNA can be determined using a range of protein concentrations. Procedures are similar to the one described above except samples were loaded on alternate lanes. Mark the four corners of the filter containing the dried gel with fluorescence dye (Scienceware high-energy autoradiography pen, cat. no. 13351) (to line up the X-ray film with the gel). Following autoradiography, bands corresponding to bound and unbound DNA are excised from the dried gel and quantified by liquid scintillation counting. Alternatively, the bands can be quantified by a phosphoimager. K_d values are obtained from a computer-fitted curve generated by the Enzfitter program *(29)*.

3.5. MutY Nicking Assay

3.5.1. Nicking Assay

The nicking activity of MutY is the combined action of the glycosylase and AP lyase activities. MutY nicks on the A-containing strand at the first phosphodiester bond (*see* **Notes 5** and **6**).

1. To each reaction, add the following in order to a microcentrifuge tube.

Sterile dH_2O	7 µL
10X MutY reaction buffer	1 µL
3' end-labeled DNA (1.8 fmol)	1 µL
MutY (72 n*M*)	1 µL
Total	10 µL

   Assay both A/G- and C/G-containing DNA. A control incubation consisting of DNA only (no MutY protein) should also be run. Incubate all reactions at 37°C for 30 min.
2. Stop the reactions in a dry ice-ethanol bath and dry the samples in a desiccator for 45 min (*see* **Note 7**).
3. Resuspend each tube in 3 µL of sequencing dye. Heat samples at 90°C for 2 min.
4. Analyze the reaction products on a 14% polyacrylamide DNA sequencing gel (IBI STS45i DNA sequencing unit), which has been prerun for more than 30 min at 1800 V.
5. Run the gel at 2000 V until bromophenol blue has migrated approximately half-way down the gel.
6. Remove the glass plates and transfer the gel onto a used X-ray film for support.
7. Cover the gel with plastic film and autoradiograph until the proper exposure is achieved. The nicked product (9 nucleotides long) migrates just below the bromophenol blue and the intact DNA (20 nucleotides long) migrates between xylene cyanol and bromophenol blue (*see* **Note 7**).

3.5.2. Kinetic Determination

Kinetic analyses are performed using a concentration range of 20-mer DNAs with fixed protein concentrations. Reactions are performed as in the nicking assay up to **step 3**, but the products are analyzed on a 14% polyacrylamide DNA sequencing gel (BRL cat. no. 21070-016 with 0.8-mm-thick gel) that has been prerun at 600 V for 30 min. Run the gel at 600 V for 40 min or until bromophenol blue has migrated approximately half-way down the gel. Remove one glass plate, put the gel in a tray, and fix the gel with 5%

acetic acid/10% methanol for 20 min. Transfer the gel onto a sheet of 3MM filter paper and dry the gel under vacuum for 1 h. Mark the four corners of the filter containing the dried gel with fluorescence dye (to line up the X-ray film with the gel). Following autoradiography, bands corresponding to cleavage products and intact DNA are excised from the gel and quantified by liquid scintillation counting. Alternatively, the cleaved product and intact DNA can be quantified by a phosphoimager. K_m and V_{max} values are obtained from a computer-fitted curve generated by the Enzfitter program *(29)*.

3.6. Formation of MutY-DNA Covalent Complex (Trapping Assay)

3.6.1. With Labeled DNA

Reactions are carried out as described in the MutY nicking assay (*see* **Subheading 3.5.1.**) except that the reactions are performed in the presence of 100 m*M* $NaBH_4$. A $NaBH_4$ stock solution (1 *M*) is freshly prepared immediately prior to use (*see* **Note 8**).

1. To each reaction, add the following in the order (it is important to add the enzyme before $NaBH_4$.) listed to a microcentrifuge tube.

Sterile dH_2O	5 µL
10X MutY reaction buffer (without NaCl)	1 µL
3' end-labeled DNA (1.8 fmol)	1 µL
MutY (72 n*M*)	1 µL
1 *M* $NaBH_4$	2 µL
Total	10 µL

2. After incubation at 37°C for 30 min, 2.5 µL of 5X SDS sample dye is added to the tube.
3. Boil the products for 2 min.
4. Load the samples on a 12% polyacrylamide gel in the presence of SDS (SDS-PAGE) according to Lemmli *(***30***)*.
5. Electrophoresis at 30 mA per gel until the dye migrates to 0.5 cm from the bottom of the gel.
6. Dry the gel onto a 3MM filter paper for 45 min and autoradiograph.

3.6.2. With Unlabeled DNA

Reactions are carried out as described in **Subheading 3.6.1.** except that the reactions (20 µL) contain 100 pmol of A/G or A/GO-containing DNA and 4 µg (about 100 pmol) of MutY. After incubation at 37°C for 30 min, 5 µL of SDS sample dye is added to the tube and the products are boiled for 2 min and separated on a 12% SDS-PAGE. The gel is stained with Coomassie blue for 1 h and then destain in 10% acetic acid for several changes.

4. Notes

1. The specific activity of the labeled DNA should be higher than 2×10^6 cpm/pmol. If the DNA substrates are not labeled well, check DNA concentration, anneal steps, Klenow fragment, and polynucleotide kinase.
2. To induce the production of MutY, the duration and temperature of induction are important. A lower temperature (20°C) is better than 37°C as the protein or mutant proteins have low solubility.
3. All enzyme purification procedures should be done at 4°C. A general precaution is to avoid bubbles in the protein samples and repeated freeze and thaw. Protease inhibitors should be included to prevent protein degradation. In addition, do not let the columns run dry. Enzymes are stored as 2-µL aliquots at –80°C and used only once. Enzyme dilution steps should be gentle: this can be done with pipeting up and down several times, gentle tapping with the fingers, or brief mixing with a vertex mixer (about 2 s) at the lowest speed.
4. If no protein–DNA complex is found, increase protein concentration, increase labeled DNA, and check the concentrations of NaCl and poly (dI-dC). When MutY homologs from different organism are assayed, find the optimal concentration of NaCl and poly(dI-dC) for the enzyme. The reaction buffer for human MYH is 10 m*M* Tris-HCl (pH 7.6), 0.5 m*M* DTT, 0.5 m*M* EDTA, and 1.5% (v/w) glycerol. Sometimes, the reactions can be enhanced by adding 50 µg/mL of BSA.
5. The ratios of protein to DNA should be more than 5 for both binding and nicking to A/G- and A/C-containing DNA substrates. For A/GO-containing DNA, the protein to DNA ratios should be 1 and 5 for binding and nicking reactions, respectively. When MutY concentration is higher than 300 n*M* in the binding reaction, multiple protein-DNA complexes can be found.
6. To obtain a clean background in the nicking assay, the oligonucleotides need to be gel purified and all solutions should be sterile.
7. In the nicking assay, if nucleases are a problem, the concentration of EDTA can be increased to 5 m*M*. Although MutY protein has been shown by Tsai-Wu et al. ***(16)*** to have both DNA glycosylase and AP lyase activities, the AP lyase activity has also been reported by others to be very weak in their MutY preparations. To complete the strand cleavage, after the MutY reaction, piperidine can be added to the sample to a final concentration of 1 *M* and then heated at 90°C for 30 min. Samples are lyophilized, resuspended with 3µL of sequencing dye, heated at 90°C for 2 min, and loaded to the gel.
8. The bottle of $NaBH_4$ should be tightly sealed after being opened and fresh solution of $NaBH_4$ should be prepared immediately prior to use. When MutY homologs from different organism are assayed, find the optimal concentration of $NaBH_4$ and pH for the enzymes.

References

1. Ames, B. N. and Gold, L. S. (1991) Endogenous mutagens and the causes of aging and cancer. *Mutat. Res.* **250,** 3–16.

2. Kasai, H. and Nishimura, S. (1991) Formation of 8-hydroxyguanine in DNA by oxygen radicals and its biological significance, in *Oxidative stress: oxidants and antioxidants* (Sies H., ed.), Academic, London, pp. 99–116.
3. Moriya, M. (1993) Single-stranded shuttle phagemid for mutagenesis studies in mammalian cells: 8-Oxoguanine in DNA induces targeted G.C to T.A transversions in simian kidney cells. *Proc. Natl. Acad. Sci. USA* **90,** 1122–1126.
4. Moriya, M., Ou, C., Bodepudi, V., et al. (1991) Site-specific mutagenesis using a gapped duplex vector: a study of translesion synthesis past 8-oxodeoxyguanosine in *Escherichia coli. Mutat. Res.* **254,** 281–288.
5. Wood, M. L., Dizdaroglu, M., Gajewski, E., and Essigmann, J. M. (1990) Mechanistic studies of ionizing radiation and oxidative mutagenesis: genetic effects of single 8-hydroxyguanine (7-hydro-8-oxoguanine) residue inserted at a unique site in a viral genome. *Biochemistry* **29,** 7024–7032.
6. Cheng, K. C., Cahill, D. S., Kasai, H., et al. (1991) 8-Hydroxyguanine, an abundant form of oxidative DNA damage, causes G-T and A-C substitutions. *J. Biol. Chem.* **267,** 166–172.
7. Michaels, M. L. and Miller, J. H. (1992) The GO system protects organisms from the mutagenic effect of the spontaneous lesion 8-hydroxyguanine (7,8-dihydro-8-oxo-guanine). *J. Bacteriol.* **174,** 6321–6325.
8. Tchou, J. and Grollman, A. P. (1993) Repair of DNA containing the oxidatively-damaged base 8-hydroxyguanine. *Mutat. Res.* **299,** 277–287.
9. Tajiri, T., Maki, H., and Sekiguchi, M. (1995) Functional cooperation of MutT, MutM and MutY proteins in preventing mutations caused by spontaneous oxidation of guanine nucleotide in *Escherichia coli. Mutat. Res.* **336,** 257–267.
10. Au, K. G., Cabrera, M., Miller, J. H., and Modrich, P. (1988) *Escherichia coli mutY* gene product is required for specific A/G to C:G mismatch correction. *Proc. Natl. Acad. Sci. USA* **85,** 9163–9166.
11. Lu, A-L. and Chang, D.-Y. (1988) Repair of single base pair transversion mismatches of *Escherichia coli in vitro*: correction of certain A/G mismatch is independent of *dam* methylation and host *mutHLS* gene functions. *Genetics* **118,** 593–600.
12. Lu, A-L. and Chang, D.-Y. (1988) A novel nucleotide excision repair for the conversion of an A/G mismatch to C/G base pair in *E. coli*. *Cell* **54,** 805–812.
13. Michaels, M. L., Cruz, C., Grollman, A. P., and Miller, J. H. (1992) Evidence that MutM and MutY combine to prevent mutations by an oxidatively damaged form of guanine in DNA. *Proc. Natl. Acad. Sci. USA* **89,** 7022–7025.
14. Radicella, J. P., Clark, E. A., and Fox, M. S. (1988) Some mismatch repair activities in *Escherichia coli*. *Proc. Natl. Acad. Sci. USA* **85,** 9674–9678.
15. Su, S.-S., Lahue, R. S., Au, K. G., and Modrich, P. (1988) Mispair specificity of methyl-directed DNA mismatch correction *in vitro*. *J. Biol. Chem.* **263,** 6829–6835.
15a. Li, X., Wright, P. M., and Lu, A.-L. (2000) The C-terminal domain of MutY glycosylase determines the 7,8-dihydro-8-oxo-guanine specificity and is crucial for mutation avoidance. J. Biol. Chem., in press.

15b. Zhang, Q.-M., Ishikawa, N., Nakahara, T., and Yonei, S. (1998) *Escherichia coli* MutY protein has a guanine-DNA glycosylase that acts on 7,8-dihydro-8-oxoguanine:guanine mispair to prevent spontaneous G:C to C:G transversions. *Nucelic Acids Res.* **26,** 4669–4675.
16. Tsai-Wu, J.-J., Liu, H.-F., and Lu, A-L. (1992) *Escherichia coli* MutY protein has both N-glycosylase and apurinic/apyrimidinic endonuclease activities on A.C and A.G mispairs. *Proc. Natl. Acad. Sci. USA* **89,** 8779–8783.
17. Lu, A-L., Yuen, D. S., and Cillo, J. (1996) Catalytic mechanism and DNA substrate recognition of *Escherichia coli* MutY protein. *J. Biol. Chem.* **27,** 24,138–24,143.
18. Manuel, R. C. and Lloyd, R. S. (1997) Cloning, overexpression, and biochemical characterization of the catalytic domain of MutY. *Biochemistry* **36,** 11,140–11,152.
19. Gogos, A., Cillo, J., Clarke, N. D., and Lu, A-L. (1996) Specific recognition of A/G and A/8-oxoG mismatches by *Escherichia coli* MutY: removal of the C-terminal domain preferentially affects A/8-oxoG recognition. *Biochemistry* **35,** 16,665–16,671.
20. Lu, A-L., Tsai-W, J.-J., and Cillo, J. (1995) DNA determinants and substrate specificities of *Escherichia coli* MutY. *J. Biol. Chem.* **270,** 23,582–23,588.
21. Yeh, Y.-C., Chang, D.-Y., Masin, J., and Lu, A-L. (1991) Two nicking enzymes systems specific for mismatch-containing DNA in nuclear extracts from human cells. *J. Biol. Chem.* **266,** 6480–6484.
22. McGoldrick, J. P., Yeh, Y.-C., Solomon, M., Essigmann, J. M., and Lu, A.-L. (1995) Characterization of a mammalian homolog of the *Escherichia coli* MutY mismatch repair protein. *Mol. Cell. Biol.* **15,** 989–996.
23. Lu, A-L. and Fawcet, W. P. (1998) Characterization of the recombinant MutY homolog, an adenine DNA glycosylase, from *Schizosacchromyces pombe. J. Biol. Chem.* **273,** 25,098–25,105.
24. Slupska, M. M., Baikalov, C., Luther, W. M., Chiang, J.-H., Wei, Y.-F., and J. H. Miller. (1996) Cloning and sequencing a human homolog (*hMYH*) of the *Escherichia coli mutY* gene whose function is required for the repair of oxidative DNA damage. *J. Bacteriol.* **178,** 3885–3892.
24a. Slupska, M. M., Luther, W. M., Chiang, J. H., Yang, H., and Miller, J. H. (1999) Functional expression of hMYH, a human homolog of the *Escherichia coli* MutY protein. *J. Bacteriol.* **181,** 6210–6213.
24b. Takao, M., Zhang, Q. M., Yonei, S., and Yasui, A. (1999) Differential subcellular localization of human MutY homolog (hMYH) and the functional activity of adenine:8-oxoguanine DNA glycosylase. *Nucleic Acids Res.* **27,** 3638–3644.
24c. Tsai-Wu, J.-J., Su, H.-T., Wu, Y.-L., Hsu, S.-M., and Wu, C. H. H. (2000) Nuclear localization of the human MutY homologue hMYH. *J. Cell. Biochem.,* in press.
25. Lu, A-L. and Hsu, I. C. (1992) Detection of single DNA base mutations with mismatch repair enzymes. *Genomics* **14,** 249–255.
26. Hsu, I.-C., Yang, Q. P., Kahng, Y. W., and Xu, J.-F. (1994) Detection of DNA point mutations with DNA mismatch repair enzymes. *Carcinogenesis* **15,** 1657–1662.

27. Maxam, A. M. and Gilbert, W. (1980) Sequenceing end-labelled DNA with base-specific chemical cleavage. *Methods Enzymol.* **65,** 499–560.
28. Bradford, M. (1976) A rapid and sensitive method for the quantitation of microgram quantities of protein utilizing the principle of protein-dye binding. *Anal. Biochem.* **72,** 248–254.
29. Leatherbarrow, R. J. (1987) *Enzfitter: A Non-linear Regression Analysis Program for IBM PC,* Elsevier Science, Amsterdam.
30. Laemmli, U. K. (1970) Cleavage of structural protein during the assembly of the head of bacteriophage T4. *Nature* **227,** 680–685.

2

Assays for the Repair of Oxidative Damage by Formamidopyrimidine Glycosylase (Fpg) and 8-Oxoguanine DNA Glycosylase (OGG-1)

Amanda J. Watson and Geoffrey P. Margison

1. Introduction

Oxidative damage produced by endogenously and exogenously generated reactive oxygen species (ROS) has been implicated in mutagenesis and carcinogenesis and may play an important role in the pathogenesis of aging *(1)*. Among ROS, the hydroxyl radical is highly reactive, producing a variety of purine- and pyrimidine-derived lesions in DNA *(2,3)*. A major pathway of hydroxyl radical-induced DNA damage involves attack on the C8 position of purines to produce 8-oxoG (7,8-dihydro-8-oxoguanine), 8-oxoA (7,8,-dihydro-8-oxoadenine) and imidazole ring fragmented lesions (formamidopyrimidines *[2,4]*). There is strong evidence to suggest that the 8-oxoG lesion, which is produced in abundance, is highly mutagenic in vitro and in vivo *(5,6)*. Such oxidized purines are primarily repaired by the base excision repair pathway, the initial step of which is excision of the modified base by DNA glycosylases *(7,8)*.

The Fpg (MutM) protein of *Escherichia coli* is a DNA glycosylase/AP lyase that efficiently releases modified purines such as 8-oxoG (when paired with cytosine in duplex DNA) and 2,6-diamino-4-hydroxy-5-*N*-methylformamidopyrimidine (me-Fapy-G *[9,10]*). The cDNA encoding the eukaryotic homolog of Fpg, OGG-1 has now been isolated by a number of laboratories *(11–15)*. OGG-1 demonstrates a similar substrate specificity to Fpg, excising 8-oxoG, preferentially when it is paired with cytosine (but being inactive when paired with adenine), and me-Fapy-G *(11,14,16)*. More recently, a second mammalian 8-oxoG-DNA glycosylase, OGG-2 has been isolated *(12,17)*, which prefers 8-oxoG paired with adenine and guanine and it has been proposed *(17)* that

From: *Methods in Molecular Biology, vol. 152: DNA Repair Protocols: Prokaryotic Systems*
Edited by: P. Vaughan © Humana Press Inc., Totowa, NJ

OGG-1 and OGG-2 have distinct anti-mutagenic functions in vivo. OGG-1 prevents mutation by removing 8-oxoG formed in DNA *in situ* and paired with cytosine, whereas OGG-2 removes 8-oxoG that is incorporated opposite adenine in DNA from ROS-induced 8-oxodGTP. Hazra et al. ***(17)*** also report, in HeLa cell extract, the presence of a protein that is an 8-oxoG- specific binding protein or an inhibitor specific for both OGG-1 and OGG-2. This protein, if found to be ubiquitously present in mammalian cells and tissue would obviously have an impact on measurement of OGG activity in crude mammalian extracts.

Two methods have been developed to assay Fpg and OGG-1 activity based on the substrate specificities of these enzymes. The method employed for several years in a number of laboratories, including ours, involves measuring the release of [^{3}H]-me-Fapy-G from a suitably treated methylated calf thymus DNA or poly (dG-dC) substrate ***(10)***. Briefly, substrate is incubated with cell-free extract for 15–60 min at 37°C, substrate DNA is ethanol precipitated and ethanol-soluble radioactivity, released by the enzyme into the supernatant, is measured by scintillation counting.

More recently, evidence suggesting that 8-oxoG plays an important role in a number of biological processes (*see* above) has generated a great deal of interest in developing assays that would specifically measure repair of this particular adduct. We describe here such an assay, which is based on the ability of Fpg and OGG-1 to remove the damaged base and to subsequently cleave at the resulting apurinic (AP) site via the AP lyase activity causing strand cleavage. Essentially, an oligo containing one 8-oxoguanine residue is labeled to high specific activity (SA) with [^{32}P], annealed to its oligo complement, incubated with cell-free extract, and the resulting cleavage products are analyzed by denaturing polyacrylamide gel electrophoresis (PAGE).

2. Materials

2.1. Preparation of Cell/Tissue-Free Extracts

1. FPG assay buffer 5X stock : 350 m*M* potassium-HEPES, 500 m*M* potassium chloride, 10 m*M* ethylenediaminetetraacetic acid (EDTA), 5 m*M* dithiothreitol (DTT), 25% v/v glycerol, pH 7.6. Aliquot and store at –20°C until use. Before use, dilute to 1 in 5 in ddH$_2$O. Dispose of excess, once thawed.
2. PBS (phosphate-buffered saline) : 0.8% NaCl, 0.02% KCl, 0.15% Na$_2$HPO$_4$, 0.02% KH$_2$PO$_4$, pH 7.4.
3. PMSF (phenylmethylsulphonyl fluoride; Sigma), 50 m*M* in 100% ethanol. Store at –20°C. Stable for at least 3 mo.
4. Leupeptin (Sigma), 10 mg/mL in ddH$_2$O. Store at 4°C. Stable for at least 1 mo.
5. Sonicator fitted with microtip probe suitable for ultrasonic disruption of cells in 1.5 mL Eppendorf tubes (*see* **Note 1**).

2.2 Protein Estimation

1. CBG (Coomassie brilliant blue G250) dye reagent concentrated (5X) stock: 780 m*M* CBG (Sigma, 75% dye content), 25% (v/v) ethanol (BDH analar), 7.4 *M* orthophosphoric acid, 0.01% Triton X-100 (v/v), 0.01% SDS (w/v). Store at 4°C in the dark for up to 1 yr.
   Before use, dilute to 1 in 5 in ddH_2O, leave at 4°C overnight then filter through 3MM chromatography paper (Whatman). Store at 4°C in the dark for up to 3 mo. Commercial reagents are available (*see* **Note 2**).
2. IBSA: 1mg/mL bovine serum albumin (BSA) in buffer I. Store at 4°C for up to 3 mo.
3. BSA (Sigma) protein standards: standards of 0.1, 0.2, 0.3, 0.4, 0.5, 0.6, 0.8, 1.0 mg/mL BSA in buffer I are prepared from IBSA. Standards are filtered (0.2 microns) and stored for up to 6 mo at 4°C.
4. Scintillation minivials.
5. Plastic spectrophotometer cuvets.
6. Multidispense pipet (*see* **Note 3**).
7. Spectrophotometer set to read at 595 nm.

2.3. DNA Estimation

1. TNE Buffer 10X stock: 100 m*M* Tris base, 10 m*M* EDTA, 2 m*M* NaCl, pH 7.4. Store at 4°C for up to 6 mo. Before use, dilute to 1 in 10 with ddH_2O and filter (0.45 microns). Store at 4°C for up to 3 mo.
2. Calf thymus DNA (Pharmacia Biotech, ultrapure): standards of 100, 200, 300, 400, 500 µg/mL in 1X TNE. Store at 4°C for up to 3 mo.
3. Hoechst 33258 (*bis*-benzamide) stock dye solution, 1mg/mL in ddH_2O. Store at 4°C in the dark for up to 6 mo.
4. TKO 100 minifluorometer and fluorometer cuvet (Hoefer, *see* **Note 4**) .

2.4. 8-Oxoguanine-DNA Glycosylase (OGG) Assay

2.4.1. Preparation of Oligo Substrate

2.4.1.1. ^{32}[P] Labeling and G25 Sephadex Purification of Oligo Substrate

1. Oligo-5' CGT TGT CAG AAG TAA OTT GGC CGC AGT GT 3'
   O = 8-oxoguanine (*see* **Note 5**).
   Prepare a 2.5 pmol/µL stock and store at –20°C for up to 3 mo.
2. T4 polynucleotide kinase (PNK) 10X buffer: 0.5 *M* Tris-HCl, pH 7.6, 100 m*M* $MgCl_2$, 100 m*M* 2-mercaptoethanol (supplied with enzyme *see* **item 3**).
3. T4 PNK (10^4 U/mL; Boehringer Mannheim, *see* **Note 6**).
4. Oligo sizing markers (8–32 bases; Pharmacia Biotech).
5. ^{32}[P]-γATP. Preferably 6000 Ci/mmol (NEN Life Science Products, 10 mCi/mL).
6. TE (STE) buffer: 10 m*M* Tris-HCl, 1 m*M* EDTA pH 8.0 (+0.1 *M* NaCl).
7. Sephadex G25 slurry: Mix 5 g Sephadex G25 (Sigma (G2550) with approx 50 mL TE buffer. Leave to swell overnight, then add fresh TE to make the slurry. Autoclave (20 min at 15 lb psi on liquid cycle) before use and store at 4°C.
8. 15-mL Falcon tubes (Falcon 2096).

9. Sterile glass wool.
10. Liquid scintillation counter.

2.4.1.2. Annealing of Labeled Oligo to Complement

1. Complementary oligo: 5' AC ACT GCG GCC AAC TTA CTT CTG ACA AC 3' (*see* **Note 5**). Prepare a stock solution (4 pmol/µL) and store at –20°C for up to 3 mo.
2. 0.5 *M* NaCl.
3. Heating block set at 94°C (*see* **Note 7**).

2.4.2. OGG Assay

1. Oligo substrate (*see* **Subheading 3.4.1.2.**).
2. Cleavage buffer 4X stock: 100 m*M* Tris-HCl pH 7.6, 200 m*M* KCl, 20 m*M* EDTA.
3. Denaturing loading buffer (LB): 80% formamide (v/v), 50 m*M* Tris-borate, 1 m*M* EDTA pH 8.0, 0.1% xylene cyanol (w/v), 0.1% bromophenol blue (w/v). Aliquot and store at 4°C.

2.4.3. PAGE Analysis of Cleavage Products

1. SequaGel Concentrate: 8.3 *M* Urea containing 25% (w/v) acrylamide: bis-acrylamide (19:1; National Diagnostics, Mensura).
2. SequaGel buffer: 8.3 *M* urea in 1.0 *M* Tris-Borate–20 m*M* EDTA Buffer, pH 8.3 (10X TBE).
3. SequaGel Diluent : 8.3 *M* Urea.
4. TEMED (N, N, N', N, N', N- Tetramethylethylenediamine).
5. 10% Ammonium persulphate (APS; w/v). 10% APS may be aliquoted and stored at –20°C until use. Dispose of excess once thawed.
6. Vertical minigel apparatus (*see* **Note 8**).
7. TBE 10X stock: 1.0 *M* Tris-Borate–20 m*M* EDTA, pH 8.3. Before use, dilute to 1 in 10 in dd H_2O.
8. Hamilton syringe (25 µL or 50 µL).
9. Disposable gel loading tips (Bio-Rad 223-9917).
10. Saran Wrap™ or similar.
11. Phosphor imager and image analysis software (*see* **Note 9**).

2.5. Fapy–DNA Glycosylase (FPG) Assay

*2.5.1. Preparation of FPG Substrate DNA (*see *Note 10)*

2.5.1.1. Deproteinization of DNA

1. Calf thymus DNA (*see* **Note 11**).
2. TE (*see* **Subheading 2.4.1.1.**).
3. Duran (or other wide-necked glass) bottles. Because of the hazards associated with this procedure (*see* **Note 12**) minimize the possibility of leakage by ensuring that the bottles have a good seal.

4. Proteinase K (Sigma).
5. Phenol equilibrated with 1 *M* Tris-HCl, pH 8.0. Prepare fresh as required. Add an equal volume of 1 *M* Tris, pH 8.0 to the phenol. Shake, allow to settle, and aspirate off as much of aqueous phase as possible. Extreme caution must be exercised when handling and disposing of phenol (*see* **Note 12**).
6. 3 *M* NaAc, pH 4.0.
7. Absolute ethanol.
8. Ether.
9. N_2 gas.
10. Water bath set at 55°C.
11. 5-mL plastic syringe.
12. Water vacuum pump aspirator.
13. 50-mL Falcon tubes (conical bottom).

2.5.1.2. Methylation of DNA

1. Duran (or other wide-necked glass) bottles.
2. 0.02 *M* Ammediol (2-Amino-2-methyl-1,3-propanediol), pH 10.0 (Sigma).
3. 5 mCi [^{3}H] MNU. Preferably approx 20 Ci/mmol (Amersham International, 1 mCi/mL in ethanol). Use immediately on delivery.
4. Absolute ethanol.
5. Pasteur pipets.
6. Water vacuum pump aspirator.
7. Ether.
8. Ethanol: Ether (1:1 v/v).
9. N_2 gas.
10. Chemical fume cupboard.

2.5.1.3. Imidazole Ring Opening of 7-Methylguanine to Generate FPG Substrate DNA

1. Sodium phosphate buffer: 50 m*M*, pH 11.4.
2. Absolute ethanol.
3. 3 *M* NaAc, pH 4.8.
4. FPG assay buffer (*see* **Subheading 2.1.**).

2.5.2. FPG Assay

1. Substrate DNA.
2. FPG assay buffer (*see* **Subheading 2.1.**).
3. FPG assay stop solution: 2 *M* sodium chloride, 1 mg/mL BSA, 0.5 mg/mL salmon sperm DNA (sonicated; *see* **Note 13**).
4. Ice-cold ethanol.
5. Scintillation minivials.
6. Aqueous scintillation cocktail (e.g., Ecoscint; National Diagnostics/Mensura).
7. Liquid scintillation counter.

3. Methods

3.1. Preparation of Extracts

Samples (tissues, cells, and extracts) must be stored on ice throughout the procedure to conserve Fpg/OGG-1 activity.

1. Preparation and storage of cells/tissues/lymphocytes.
   Cells (bacterial or mammalian): harvest, wash with PBS, store cell pellet at –20°C.
   Tissue: snap-freeze (dry ice or liquid nitrogen) and store at –20°C (<1 mo) or –70°C (>1 mo).
   Lymphocytes: collect whole blood into universal containing EDTA (final concentration 25 m*M*). Isolate lymphocytes by density centrifugation ***(18)***, wash with PBS, and store cell pellet at –20°C.
   The number of cells or weight of tissue required for the assay will depend on the level of activity (e.g., *see* **Note 14**).
2. Transfer tissue or cell pellet to 1.5-mL Eppendorf tube in ice and add cold buffer I (500–1000 µL) containing 5 µg/mL leupeptin (*see* **Note 15**).
3. Sonicate sample (*see* **Note 1**) within a MSC class I cabinet to minimize exposure to aerosols. It may be necessary to mince with fine scissors or add glass beads to the sample to aid ultrasonic disruption (*see* **Note 16**).
4. Add PMSF (50 m*M* solution in alcohol) to the sample immediately following sonication so that the final concentration is 0.5 m*M* (i.e., 1/100 of volume).
5. Centrifuge at 15,000–20,000*g* for 10 min at 4°C (*see* **Note 17**).
6. Transfer supernatant to a clean Eppendorf tube in ice. Extracts are now ready for use. For short-term storage (≤ = 48 h), in ice preferably in cold room/cabinet, is recommended. If for longer periods, extracts should be snap-frozen (dry ice or liquid nitrogen) and stored at –20°C. Activity may be lost on freeze-thawing, but we have not systematically investigated this.

3.2. Protein Estimation (19) (see Note 18)

1. Switch on the spectrophotometer at least 10 min before taking readings so that the lamp is ready.
2. Add, in duplicate, 40 µL of each BSA standard or unknown (*see* **Note 19**) to the bottom of a scintillation minivial. Blank tubes contain buffer I only.
3. Add 2 mL of CBG/ Bio-Rad reagent to each tube and gently mix (*see* **Note 3**).
4. Transfer 1 mL of blank sample to the cuvet and zero the machine at 595 nm. Repeat with duplicate blank sample to check reproducibility and stability of readings (*see* **Note 20**).
5. Transfer 1 mL of the lowest standard to the cuvet, record the reading, empty, and drain the cuvet thoroughly by blotting upside down on a paper towel and repeat with the next standard sample.
6. After reading all standard samples, transfer 1 mL of the unknown sample to the cuvet, record the reading, empty, and drain the cuvet thoroughly by blotting

upside down on a paper towel and repeat with the next unknown sample (*see* **Note 21**).

7. Construct a standard curve by plotting absorbance of standards versus protein concentration (mg/mL). Calculate mean protein concentration (mg/mL) of duplicate samples (*see* **Note 21**) by reference to the standard curve.

3.3. DNA Estimation (20,21)*(see Note 4)*

1. Switch on the TKO 100 at least 15 min before taking measurements so that the lamp is ready and the temperature in the chamber stabilizes.
2. Freshly prepare working solution of Hoechst 33258 by diluting the stock solution to 1 μg/mL in 1X TNE, wrap in foil to protect from light, and allow to warm to room temperature before use.
3. Set the sensitivity of the detector monitor to about 50% by turning the scale knob approx 5 full clockwise turns from the fully counter position.
4. Add 2 mL of Hoechst working solution to the cuvet, if necessary wipe the sides of the cuvet with a low-lint tissue and place in the sample chamber.
5. Zero the reading.
6. Deliver 2 μL of lowest DNA standard into the 2-mL dye solution and mix by pipetting the solution into and out of a disposable pipet several times without introducing bubbles.
7. Close the sample chamber and record the reading (*see* **Note 22**).
8. Remove the cuvet from the sample chamber. Empty and drain the cuvet thoroughly by blotting it upside down on a paper towel between readings.
9. Repeat **steps 4–8** at least once to verify that results are reproducible (*see* **Note 21**).
10. Repeat **steps 4–9** with rest of DNA standards.
11. Finally, repeat steps 4-9 with unknown sample (*see* **Note 23**).
12. Construct a standard curve by plotting standard readings vs DNA concentration (μg/mL) of standards. Calculate mean DNA concentration (μg/mL) of replicate unknown samples by reference to the standard curve.

3.4. 8-Oxoguanine-DNA Glycosylase (OGG) Assay

Before starting work, ensure that appropriate shielding is in place and that a radiation monitor is at hand to monitor for possible contamination. Handling and disposal of [^{32}P] must be performed in accordance with local rules pertaining to radioactive substances.

3.4.1. Preparation of Oligo Substrate

3.4.1.1 [^{32}P] Labeling and Sephadex G25 Purification of Oligo Substrate

1. Set up the following reaction in an Eppendorf tube:
   1 μL oligo (2.5 pmol) or 1 μL oligo sizing marker
   1 μL 10X PNK buffer
   3.8 μL ddH_2O

Working behind a perspex screen, add:
4 µL ^{32}[P]- γATP
0.2 µL PNK

2. Incubate reactions at 37°C for 30 min. During this time, prepare the Sephadex G25 columns.
3. At the end of the incubation, add 40 µL TE to the mix, transfer 1 µL to a scintillation minivial, add 2 mL scintillation cocktail and count (*see* **Note 24**). Proceed to purify the rest by passing through a G25 column.
4. Pack the tip of a 1mL plastic disposable syringe with sterile glass wool. Use the plunger to pack the wool tightly to approx 50µL.
5. Pipet 1 mL of the Sephadex/TE slurry to the syringe, place the syringe in a 15-mL Falcon tube (to act as a carrier in the rotor), and centrifuge at 1700*g* for 4 min at room temperature.
6. Repeat **step 2** until a packed column bed of 0.8–0.9 mL is obtained.
7. Wash and equilibrate the column by applying 100 µL STE to the top of the column and centrifuging at 1700*g* for 4 min at room temperature.
8. Place column in a fresh Falcon tube and apply the oligo (50 µL) to the top of the column.
9. Centrifuge at 1700*g* for 4 min at room temperature and transfer the eluate to a 1.5-mL Eppendorf. Dispose of column in accordance with local rules pertaining to radioactive substances.
10. Transfer 1 µL of eluate to a scintillation minivial, add 2 mL scintillation cocktail and count.
11. Calculate percent incorporation of label (*see* **Note 24**).

3.4.1.2. Annealing of Labeled Oligo to Complement

1. Remove 40 µL of the column purified oligo (approx 2 pmol) into an Eppendorf tube, add twofold excess of complementary oligo (i.e., 4 pmol), 5 µL 0.5 *M* NaCl and ddH_2O to a total volume of 50 µL (*see* **Note 25**).
2. Place tube in a heating block (*see* **Note 7**) set at 94°C, incubate for 2–3 min then switch off block and allow tube to cool to room temperature slowly.
3. Efficiency of annealing may be checked by PAGE (*see* **Note 26**).

3.4.2. OGG Assay

1. Set up the following reaction (*see* **Note 27**) in an Eppendorf tube in ice:
   1 µL labeled, annealed oligo
   2.5 µL 4X cleavage buffer
   0–6.5 µL cell/tissue extract
   and add an appropriate volume of ddH_2O to give final volume of 10 µL.
2. Vortex mix and centrifuge briefly to collect tube contents.
3. Incubate at 37°C for 15–60 mins (*see* **Note 28**).
4. Add 20 µL denaturing LB and either proceed to next stage (analysis of cleavage products) or store at –20°C until required (*see* **Note 29**).

3.4.3. PAGE Analysis of Cleavage Products

1. Wash all minigel apparatus carefully with ddH_2O and dry thoroughly (*see* **Note 30**).
2. Assemble minigel apparatus according to manufacturer's instructions.
3. Prepare 20% SequaGel by mixing 8 mL SequaGel Concentrate (25%) with 1 mL SequaGel Buffer and 1 mL SequaGel Diluent. Just prior to pouring add 100 µL 10% APS, mix well, then immediately add 10 µL TEMED and mix. Pipet into apparatus avoiding air bubbles.
4. Following polymerization, remove comb and, using a Hamilton syringe, rinse wells thoroughly (*see* **Note 31**) with running buffer (i.e., TBE).
5. Heat samples for 5 min at 95°C and cool on ice for 10 min.
6. Rinse each well with 1X TBE just before loading 6 µL of each sample, preferably using disposable tips. Load labeled size marker (*see* **Subheading 3.4.1.1.**) in order to confirm size of cleaved products.
7. Run gel in 1X TBE at 100–200 V until bromophenol blue dye front is approx 1 cm from the bottom of the gel.
8. At the end of the run remove one plate and wrap the gel (still supported on the other plate) with Saran Wrap. Detect bands using phosphorimager or autoradiography, confirming size of products by reference to the size marker. Calculate percent oligo cleaved by image analysis (*see* **Note 9**).

3.4.4. Calculation of Enzyme Activity

1. Calculate nmoles oligo cleaved in each reaction by multiplying % oligo cleaved (determined by image analysis) by amount of oligo (in nmoles) in reaction.
2. Plot nmoles oligonucleotide cleaved vs mg protein/ µg DNA in extract and from the linear part of the curve calculate nmoles oligo cleaved/mg protein or /µg DNA, respectively.
3. Divide nmoles oligo cleaved/mg protein or /µg DNA by incubation time in hours to give specific activity in nmoles oligo cleaved /mg protein/h or /µg DNA/h.

3.5. Fapy-DNA Glycosylase (FPG) Assay

3.5.1. Preparation of FPG Substrate DNA

3.5.1.1. Deproteinization of Substrate DNA

1. Dissolve CT DNA at 2 mg/mL in TE (up to 300 mL) on a stirrer overnight in a IL Duran bottle. There will be some insoluble bits, but it is not necessary to remove them.
2. Place bottle in 55°C water bath for 5 min then add solid proteinase K (1 mg/10 mL DNA solution). The bits should disappear quickly, but leave for 1 h swirling occasionally before adding another 1 mg of proteinase K per 10-mL solution.
3. After a further 1 h at 55°C, move to fume cupboard on tray, cool under running tap water, and add equal volume of phenol equilibrated to pH 8.0 using 1 *M* Tris (*see* **Subheading 2.5.1.1.**). Cap and shake vigorously for 5 min—be aware of the possibility of leakage.

4. Allow to stand for about 1.5 h at room temperature: the phenol should settle out and can be almost completely removed by aspiration through the upper aqueous layer using a glass pipet.
5. Decant supernatant into 50-mL Falcon tubes (conical bottom) and spin at 1000*g*, room temperature, 10 min.
6. Observe interface carefully: if clear, re-extraction is not necessary (*see* **Note 32**). Remove all traces of phenol from bottom of tube using glass 5-mL pipet or Pasteur pipet and rubber pipet bulb or pipet pump. Do not worry about taking some of the aqueous layer. Pour off supernatants into Duran bottle of appropriate capacity.
7. Add 1/10 vol of 3*M* NaAc pH 4.0 to pooled aqueous phases, mix well, and add 2 vol cold ethanol. Cap and mix by inversion.
8. Lift out DNA on glass pipet and transfer to smaller Duran. Wash three times with ethanol at room temperature by vigorous shaking and water vacuum pump aspiration of the ethanol. Make sure DNA spreads out in ethanol to ensure complete penetration of ethanol.
9. Wash at room temperature three times with ethanol:ether (1:1) and then three times with ether alone. Each time, pour off the washes into a tray in a fume cupboard for evaporation (no naked flames/electrical appliances) or alternatively dispose of according to local rules.
10. Dry DNA in stream of N_2 to remove the ether, teasing apart fibrous DNA with Pasteur pipets. Dry to constant weight.

3.5.1.2. Methylation of DNA

Because of the radiochemical hazard involved, the following procedure should be carried out in a fume cupboard with an appropriate airflow rate. Handling and disposal of [3H] must be performed in accordance with local rules pertaining to radioactive substances. We advise monitoring for [3H] contamination of the work area before starting and, of course, on completion.

1. For 5 mCi [^{3}H]-MNU in 1 mL ethanol: dissolve 40 mg DNA on a stirrer plate overnight at 8 mg/mL in 0.02 *M* Ammediol, pH 10.0 in a 25 mL Duran bottle. Transfer 2 mL of this solution to a separate container (This is to be used for rinsing [^{3}H] vial—*see* **step 3**).
2. In tray in fume cupboard, CAREFULLY remove seal from [^{3}H]MNU vial using blunt forceps. Use a twisting rather than pulling action and put aluminium ring and sealing disk directly into beaker in tray. Recap vial with black plastic cap provided—avoid shaking.
3. Put Duran on stirrer in tray, and using 5-mL plastic syringe, carefully transfer MNU solution into stirring DNA solution. Rinse out vial with two 1-mL aliquots of DNA solution by serial transfer. Put syringe and empty vial in sink for careful rinsing.
4. Continue stirring DNA for 5 h at room temperature. Carefully remove stirrer bar, then add 1/10 volume of 3 *M* NaAc, pH 4.0 and 2 vol of cold ethanol. Form DNA precipitate by swirling and inversion being very careful of leakage—any spills will contain [^{3}H]-methanol, which will blow off rapidly.

5. When DNA has formed a tight ball, carefully remove supernatant using a Pasteur pipet attached to a water vacuum pump, the outflow of which is piped directly into the sink drain hole to avoid splashing. Do not be concerned about tiny fragments of DNA being sucked down the sink, but do avoid the bulk of the DNA! It is possible to remove all of the supernatant safely in this way.
6. Wash the DNA with ethanol (about 20–30 mL per wash) making sure pellet is "fluffed" out each time. After 10 washes, check [^{3}H] radioactivity counts in 500 µL of wash (+3 mL scintillant) to monitor the washing efficiency. Ideally, the last wash should be close to background, but as small fragments of DNA may be present, two consecutive washes with similar cpm is acceptable.
7. Dry DNA by washing in ethanol:ether (1:1) twice and ether twice. DO NOT aspirate supernatants down the sink! Pour supernatants carefully into a stainless steel tray for evaporation and thorough washing down the sink.
8. Blow off residual ether in a slow stream of N_2 gas, teasing apart DNA with Pasteur pipets if necessary. Transfer to preweighed clean glass vial and dry to constant weight.

3.5.1.3. Imidazole Ring Opening of 7-Methylguanine to Generate FPG Substrate DNA

1. Dissolve DNA (2 mg/mL) in 50 m*M* sodium phosphate buffer (pH 11.4) and leave at room temperature for 24 h to open the imidazole rings of the 7 meG residues.
2. Add 0.1 vol 3 *M* NaAc (pH 4.8) and 2 vol of cold ethanol.
3. Form DNA precipitate by swirling and inversion.
4. Spool DNA, transfer to a fresh tube, and wash extensively with 70% ethanol.
5. Resuspend DNA at 2 mg/mL in 1X FPG assay buffer (without DTT) and store at 4°C.

3.5.2. FPG Assay

1. Dilute substrate with FPG assay buffer (containing DTT) to give approx 10^5 cpm/mL.
2. Into 1.5-mL Eppendorf tubes in ice, add 1–20 µL of tissue or cell extract (*see* **Subheading 3.1.**), 10 µL of 5X FPG assay buffer and make volume up to 50 µL with ddH_2O (*see* **Note 33**).
3. Start the assay reaction by adding 50 µL of diluted substrate (approx 5×10^3 cpm).
4. Vortex briefly, and centrifuge to collect contents in the bottom of the tubes.
5. Incubate tubes at 37°C for 15–60 min (*see* **Note 28**).
6. At the end of this time, add 25 µL of FPG assay stop solution with vortex mixing, followed immediately by 250 µL of ice-cold ethanol.
7. After cooling the tubes for 20 min on dry ice (or at –80°C), centrifuge at 25,000*g* for15 min at 4°C (see **Note 34**).
8. Carefully remove 300 µL of supernatant into scintillation minivials (*see* **Note 35**), mix with 3 mL Ecoscint and count for 5 min in a suitable scintillation counter.
9. Calculate the cpm by averaging two 5-min counts. Unusually high counts, particularly in the first few tubes, may indicate chemiluminescence and these tubes should be recounted to constant cpm.

3.5.3. Calculation of Enzyme Activity

1. Calculate total cpm released from each test sample by subtracting mean of "blank tubes" (*see* **Note 33**) and multiplying by 380/300 (*see* **Note 36**).
2. Plot total cpm vs μg protein or ng DNA and from the linear part of the curve calculate cpm/μg protein of cpm/ng DNA.
3. Multiply cpm/μg protein or cpm/ng DNA by the conversion factor (*see* **Note 37**) to give FPG specific activity in pmoles Fapy released/g protein/h or pmoles Fapy released/mg DNA/h, respectively.

4. Notes

1. We routinely use a Heatmaster Sonicator fitted with a microtip probe (standard tapered with 3.2-mm diameter) and sonicate for 10 s at 216 μm (peak-to-peak amplitude). Appropriate sonication conditions will depend upon the particular machine and probe used and should be established by experiment.
2. Bio-Rad dye reagent (Bio-Rad) supplied as a 5X stock may be used as an alternative to CBG. Standard curves using CBG are linear up to 1 mg/mL compared to those using Bio-Rad, which plateau around 0.6–0.8 mg/mL. We consider absorbances less than half that of the lowest standard as inaccurate, and absorbances between 0.1 and 0.6 to be ideal.
3. We recommend using an Eppendorf multipet plus (Eppendorf) or similar positive displacement machine to repeat dispense quickly and accurately.
4. Hoechst 33258 binds to the minor groove of DNA. When 365 nm light (long UV) excites this dye, fluorescence results and thus can be measured by minifluorometer (as described here), fluorescence spectrophotometer ***(21)*** or fluorescence microtiter plate-reader. We have also used a microtiter (96-well) based assay;10 μL sample or standard (range 0–50 μL/mL DNA) and 100 μL Hoechst (1 μL/mL) per well.
5. We purchase our 8-oxoguanine-containing oligos (synthesized according to standard phosphoramidite chemistry) from either Alta Bioscience, School of Biochemistry, University of Birmingham, UK or Genset Oligos, Paris, France. Nonmodified oligos are now readily available at a reasonable price from a number of commercial sources (e.g., Life Technologies and Cruachem).
6. We purchase our T4 polynucleotide kinase (PNK) from either Boehringer Mannheim or United States Biochemical (USB).
7. The annealing reactions may be heated in a PCR machine, however, it is important to allow the reactions to cool slowly to room temperature.
8. We use the Bio-Rad mini-PROTEAN II electrophoresis system.
9. We use the STORM phoshorimager (Molecular Dynamics) in combination with Imagequant software to analyze our results. Alternatively, autoradiography in combination with a suitable densitometer may be used.
10. An alternative substrate for this assay is poly (dG-dC) treated with [^{3}H]-DMS ***(10)***.
11. We routinely use Sigma calf thymus DNA (<3% protein).
12. Phenol is a powerful systemic poison, which is readily absorbed through the skin and causes chemical burns on contact. Phenol should be neutralized with an equal

volume of 4 *M* NaOH, before disposal down the sink in high dilution. Alternatively, follow local rules governing disposal of phenol.

13. Prepare this solution by mixing 8 parts 5 *M* NaCl, 2 parts 10 mg/mL BSA, 1 part 10 mg/mL salmon sperm DNA (sonicated) with 9 parts ddH_2O.
14. Cells and tissues vary in Fpg/OGG-1 activity and therefore the amount of extract required to detect activity will also vary. The following may be used as a starting guide if Fpg/OGG-1 activity is unknown.

Cells:	Bacterial	Pelleted from 1.5 mL of stationary culture
	Mammalian	10^7 cells
	Lymphocytes	10 mLs whole blood
Tissue:	1–10 mg	

More (or less) cells or tissue may be used providing sonication is effective without compromising enzyme activity.

Inclusion of a positive control (i.e., Fpg/OGG-1 expressing cell line or tissue) in both extraction and assay procedures is strongly advised, particularly when assaying samples with very low or unknown activity.

If substrate-limiting conditions are attained at the lowest volume of extract used, repeat the assay using appropriate dilutions of extract .

15. A minimum buffer volume of 500 µL is needed to avoid heating (>40°C) of the sample during sonication. If smaller extract volumes (300–500 µL) are required, it may be necessary to sonicate the sample for shorter times or on ice. Establish appropriate conditions by experiment. We find that a thermocouple thermometer is optimal for monitoring temperature during procedures.
16. Some tissues (e.g., gut and tumor) are more difficult to disrupt and therefore require the addition of 0.1–0.2 *g* glass beads (100 mesh, BDH) and/or mincing before sonication.
17. It may be possible to omit this step if very few cells are used but this should be checked by experiment.
18. If a platereader with appropriate filter is available this procedure may be adapted as a microtiter (96-well) plate assay. We have successfully used, in triplicate, 40 µL of sample or standard (linear range 0–0.1 mg/mL protein) and 200 µL of dye reagent per well.
19. Samples (e.g., liver or tumor tissue) may need to be diluted in buffer I to achieve absorbance within linear range of the standard curve.
20. Rezero the machine where necessary; frequency will depend on the particular machine.
21. Replicate protein and DNA estimations should be within 5%.
22. The observed fluorescence stabilizes after a few seconds and then begins to drop as the chamber warms.
23. Samples (e.g., tumor tissue) may need to be diluted in FPG assay buffer to achieve fluorescence within the range of the standard curve. If the reading is lower than the lowest standard, use a larger volume of extract (up to 10 µL) and apply a correction factor.

24. Percent incorporation of label = (total postcolumn cpm) / (total precolumn cpm) × 100.
To ensure that the labeled oligo has a high specific activity, the number of pmoles of [^{32}P]-γATP in the labeling reaction should be at least twice the number of pmol of oligo 5' ends. In this reaction we use 2.5 pmol oligo and 6.7 pmol of [^{32}P]-γATP. Therefore, if all the oligo is labeled, 37% of the [^{32}P] would be incorporated.
The specific activity (SA; in Ci/mmol) of the labeled oligonucleotide may be calculated as follows:

$$dpm = cpm/CE$$

where CE = counting efficiency of scintillation counter. dpm/pmol oligo = total cpm (i.e., in 50 μL) / (2.5 × CE) dpm/mmol oligo = (cpm × 10^9) / (2.5 × CE) 1 Ci = 2.22 × 10^{12} dpm therefore, SA (Ci/mmol) = (cpm × $10^{9)}$) / (2.5 × CE × 2.22 × 10^{12}).
25. To ensure efficient annealing it is important to use small reaction volumes and incubate with excess complement. We have found twofold excess of complement to be sufficient and greater excess should be avoided as this may cause problems when trying to denature the annealed oligo.
26. Annealed oligo may be run on a 20% nondenaturing gel to check efficiency of annealing. Follow the procedure described in **Subheading 3.4.3.** with three exceptions. (1) Use a nondenaturing acrylamide gel mix (e.g., Accugel, National Diagnostics) to prepare the gel, (2) Use a nondenaturing loading buffer : 0.25% bromophenol blue, 0.25% xylene cyanol, 30% glycerol, adding 2 μL loading buffer to 4 μL sample, and, (3) do not heat the samples prior to loading.
27. It is advisable to include OGG-1/Fpg expressing and nonexpressing extracts as controls in this assay. It may be necessary to increase substrate concentration when assaying OGG activity in human/mammalian cell/tissue extracts because of the presence of an 8-oxoG specific-binding protein (***17***).
28. Incubation time may be reduced for highly active extracts, however incubation times of > 1 h are not recommended: in our hands, loss of activity becomes a problem.
29. The $T_{1/2}$ for decay of ^{32}[P] is 2 wk. Therefore, it is advisable to use the substrate quickly (i.e., within 2 wk) so as not to compromise the specific activity.
30. It is very important to use clean, dry apparatus. Remains of polymerized gel can cause the plates to crack and may compromise the running of the gel. Also examine the edges of the plates carefully for chips and cracks because if an inadequate seal is formed between the bottom of the plates and the gasket the gel will leak.
31. It is essential to rinse the wells thoroughly as soon as the comb is removed so as to remove small amounts of acrylamide solution trapped by the comb, which will polymerize in the wells, and produce irregular wells that give rise to distorted bands. It is equally important to wash away excess urea, a component of the denaturing gel mix, which leaches out of the gel into the wells and affects loading if not dispersed.
32. If the interface is not clear, reextraction is necessary. Remove upper aqueous layer from each tube and combine in a Duran bottle. Repeat **steps 3–5**.

33. It is necessary to include a number of "blank" tubes to assess the counts released from the substrate in the absence of protein. The mean value of these is deducted from the counts released from each test sample. It is also advisable to include OGG-1/Fpg expressing and nonexpressing extracts.
34. The samples should be centrifuged with the brake off to avoid disruption of the pellet.
35. If it is sometimes difficult to remove 300 µL supernatant without disturbing the pellet, a smaller aliquot can be removed, but remember to take this into account when calculating the total cpm released (*see* **Note 37**).
36. Because only 300 µL out of total reaction of 380 µL are counted, cpm must be multiplied by 380/300 to give total cpm released per sample. If less supernatant is removed, total counts = cpm × 380/vol counted.
37. Conversion factor (CF) is calculated from the counting efficiency and substrate specific activity.

    (a) dpm = cpm/CE.

    where CE = counting efficiency of scintillation counting.

    (b) Substrate specific activity (SA) in Ci/mmol (= nCi/pmol). Recalculate monthly by reference to decay chart.

    (c) 1 Ci = 2.22×10^{12} dpm, therefore, 1 nCi = 2220 dpm.

    From (a), (b), and (c):

    FPG activity (in pmol Fapy released /µg /h) = cpm/µg × 1/ (SA × 2220 × CE × h).

    (Where h = assay incubation time in hours).

    Therefore,

    FPG activity (in pmol Fapy released/g/h) = cpm/µg × 10^6/(SA × 2220 × CE × h

    = cpm/µg × CF

    therefore, CF = 10^6/(SA × 2220 × CE × h).

References

1. Halliwell, B. and Gutteridge, J. M. C. (1990) Role of free radicals and catalytic metal ions in human disease: an overview. *Methods Enzymol.* **186,** 1–85.
2. Dizdaroglu, M. (1991). Chemical determination of free radical-induced DNA damage to DNA. *Free Radical Biol. Med.* **10,** 225–242.
3. Boiteux, S., Gajewski, E., Laval, J., and Dizdaroglu, M. (1992) Substrate specificity of the *Escherichia coli* Fpg protein (formamidopyrimidine-DNA glycosylase): excision of purine lesions in DNA produced by ionizing radiation or photosensitization. *Biochemistry* **31,** 106–110.
4. Cadet, J., Berger, M., Douki, T., and Ravanat, J.L. (1997) Oxidative damage to DNA: formation, measurement and biological significance. *Rev. Physiol. Biochem. Pharmacol.* **131,** 1–87.
5. Michaels, L. M. and Miller, J. H. (1992) The GO system protects organisms from the mutagenic effects of the spontaneous lesion 8-hydroxyguanine (7,8-dihydro-8-oxoguanine). *J. Bacteriol.* **1,** 6321–6325.
6. Grollman, A. P. and Moriya, M. (1993) Mutagenesis by 8-oxoguanine: an enemy within. *Trends Genet.* **9,** 246-–249.

7. Demple, B. and Harrison, L. (1994). Repair of oxidative damage to DNA: enzymology and biology. *Annu. Rev. Biochem.* **63,** 915–948.
8. Boiteux, S. and Laval, J. (1997) Repair of oxidised purines in DNA, in *Base Excision Repair of DNA Damage* (Hickson, I. D., ed.), Landes Bioscience-Springer, Austin, TX, pp. 31–44.
9. Karakaya, A., Jaruga, P., Bohr, V. A., et al. (1997) Kinetics of excision of purine lesions from DNA by *Escherichia coli* FPG protein. *Nucleic Acids Res.* **25,** 474–479.
10. Bjørås, M., Luna, L., Johnsen, B., et al. (1997) Opposite base-dependent reactions of a human base excision repair enzyme on DNA containing 7,8-dihydro-8-oxoguanine and abasic sites. *Embo J.* **16,** 6314–6322.
11. Van der Kemp, P. A., Thomas, D., Barbey, R., et al. (1996) Cloning and expression in *Escherichia coli* of the *OGG1* gene of *Saccaromyces cerevisiae*, which coded for a DNA glycosylase that excises 7,8-dihydro-8-oxoguanine and 2,6-diamino-4-hydroxy-5-methylformamidopyrimidine. *Proc. Natl. Acad. Sci. USA* **93,** 5197–5202.
12. Nash, H. M, Bruner, S. D., Scharer, O. D., et al. (1996) Cloning of a yeast 8-oxoguanine DNA glycosylase reveals the existence of a base-excision DNA-repair protein superfamily. *Curr. Biol.* **6,** 969–980.
13. Roldan-Arjona, T., Wei, Y.F, Carter-K. C., et al. (1997) Molecular cloning and functional expression of a human cDNA encoding the antimutator enzyme 8-hydroxyguanine-DNA glycosylase. *Proc. Natl. Acad. Sci. USA* **94,** 8016–8020.
14. Rosenquist, T. A., Zharkov, D. O., and Grollman A. P. (1997). Cloning and characterisation of a mammalian 8-oxoguanine DNA glycosylase. *Proc. Natl. Acad. Sci. USA* **94,** 7429–7434.
15. Tani, M., Shinmura, K., Kohno, T., et al. (1998) Genomic structure and chromosomal localisation of the mouse *Ogg1* gene that is involved in the repair of 8-hydroxyguanine in DNA damage. *Mamm. Genome* **9,** 32–37.
16. Girard, P. M., D'Ham, C., Cadet, J., and Boiteux S. (1998). Opposite base-dependent excision of 7,8-dihydro-8-oxoadenine by the ogg1 protein of *Saccharomyces cerevisiae. Carcinogenesis* **19,** 1299–1305.
17. Hazra, T. K., Izumi, T., Maidt, L., et al. (1998) The presence of two distinct 8-oxoguanine repair enzymes in human cells: their potential complementary roles in preventing mutation. *Nucleic Acids Res.* **26,** 5116–5122.
18. Coutinho, L. H., Gilleece, M. H., DeWynter, E. A., Will, A. and Testa, N. G. (1993). Clonal and long-term culture using human bone-marrow. in *Haemopoesis, A Practical Approach* (Testa, N. G. and Molyneux, G. M., eds.) Oxford University Press, Oxford, UK, p. 288.
19. Bradford, M. M. (1976) A rapid and sensitive method for the quantitation of microgram quantities of protein utilising the principle of protein-dye binding. *Anal. Biochem.* **72,** 248–272.
20. Instructions—TKO 102 Standards Kit (Hoefer Scientific Instruments UK, Newcastle-under-Lyme, UK.
21. Cesarone, C. F., Bolognesi L., and Santi L. (1979) Improved microfluorimetric DNA determination in biological material using 33258 Hoechst. *Anal. Biol.* **100,** 188.

3

Detection and Quantitation of Uracil DNA Glycosylase Activity

Geraldine M. O Grady

1. Introduction

One of the main determining factors for maintaining the informational integrity of the DNA genomes in all organisms is the efficiency of repair of DNA lesions. DNArepair mechanisms have evolved to counteract the deleterious effects of DNA damage. One such repair mechanism is Base Excision Repair (BER). BER is a repair process initiated by a class of enzymes called glycosylases. These enzymes catalyze the hydrolysis of the N-glycosidic bond, thereby liberating the damaged or inappropriate base and generating an abasic site, which is subsequently acted upon by apurinic/apyrimidinic endonucleases (AP endonucleases). Repair synthesis is then completed by DNA polymerase and DNA ligase (reviewed in **ref. *1***).

One of the most abundant and best characterized of all glycosylases is uracil DNA glycosylase (UDG) *(2)*. UDG catalyses the cleavage of the N-glycosidic bond that joins the uracil base to the deoxyribose phosphate backbone of DNA. The uracil base, whether as a result of misincorporation during DNA synthesis or deamination of cytosine (chemical or spontaneous), is recognized by this enzyme. UDG was originally detected in *Escherichia coli* *(2)* and purified to homogenity *(3)* and has since been identified in all kingdoms.

The native *E. coli* enzyme (encoded by the *ung* gene) is a 24 kDa monomeric protein that specifically releases uracil from single- or double-stranded DNA, the rate of hydrolysis on single stranded uracil DNA being twice that of duplex DNA *(3)*. The exquisite specificity of UDG is illustrated by its failure to recognise dUMP, dUTP, RNA, or uracil residues at the 3' terminal position of oligonucleotides. Activity does not require a divalent cation, occurs in the presence of ethylenediaminetetraacetic acid (EDTA) and is inhibited by free

From: *Methods in Molecular Biology, vol. 152: DNA Repair Protocols: Prokaryotic Systems*
Edited by: P. Vaughan © Humana Press Inc., Totowa, NJ

uracil and AP-site containing DNA. *E. coli* UDG is not inhibited by sulphydryl blocking agents (N-ethylmalemide and iodoacetate), which suggest that the single-cysteine residue it contains may not be involved in the active site *(3)*. However, activity of the *E. coli* enzyme is inhibited by *B. subtilis* bacteriophage PBS2 inhibitor (Ugi), an 18.5 kDa heat-stable monomeric protein that forms a stable protein–protein interaction with UDG *(4)*. UDGs from diverse biological sources are inhibited by Ugi, suggesting a highly conserved tertiary structure between all UDGs *(5)*. The biological consequences of uracil residues in DNA may have cytotoxic or lethal effects. Accumulation of uracil in DNA is potentially mutagenic and may perturb many molecular events, e.g., protein–nucleic acid interactions. It is not surprising, therefore, that the uracil-DNA repair pathway of almost every organism is remarkably similar and that UDG is one of the most highly conserved polypeptides yet identified. In this chapter, the general strategies employed for assaying for UDG activity are described.

A relatively simple approach to demonstrating UDG activity exploits the fact that bacteriophage grown on *E. coli dut* (dUTPase) *ung* cells incorporate as much as 30% uracil in place of thymine. Plasmid DNAs prepared in these bacterial hosts readily incorporate and retain the uracil base as they are replicated and these plasmids can subsequently be used as substrates in UDG assays. Incubation of plasmid substrates with UDG generates abasic sites in the plasmid, which are susceptible to thermal cleavage by alkali or enzymatically by AP endonucleases (*see* **Note 1**). Digestion products are visualized by agarose gel electrophoresis followed by ethidium bromide staining. A drawback of this assay, however, is the fact that the UDG activity is not readily quantitated, a prerequisite for any enzymatic assay. A more quantitative approach to assaying for UDG activity employs radiolabeled DNA substrates where the release of the radiolabeled uracil base is measured. A number of approaches may be taken to generate radiolabeled UDG substrates in which the uracil bases (or a proportion of them) are tritiated. Many investigators have employed the *B. subtilis* phage PBS1 whose genome naturally contains uracil in place of thymine. PBS1 tritiated DNA is prepared by infecting bacterial cells with PBS1 followed by media supplementation with tritiated uridine. Phage recovered from these cells after lysis are concentrated and viral DNA extracted, quantitated, and specific activity determined. Radiolabeled polynucleotides may also be generated in vitro by nick translation of suitable templates where tritiated dUTP is incorporated during polymerization. Here, we describe a simple, straightforward approach to assaying for UDG activity. This assay employs a PCR-generated [^{3}H]-labeled dsDNA substrate and measures the release of the [^{3}H]-uracil bases that are soluble in cold acidic solution. We also describe the less sensitive, yet simple, assay for UDG activity that employs unlabeled uracil-containing plasmid DNAs as substrates for UDG.

2. Materials

2.1. [3H]-Uracil-Containing DNA Preparation

1 ^{3}H-dUTP (Deoxy [5-^{3}H] uridine 5'-triphosphate ammonium salt, from Amersham TRK351, 1 mCi/mL, 15–20 Ci/mmol) (*see* **Note 2**).
2. PCR reagents:
   a. 10X *Taq* DNA polymerase reaction buffer: 100 m*M* Tris-HCl, pH 8.3, 500 m*M* KCL, 0.1% gelatin.
   b. 25 m*M* magnesium chloride.
   c. 10X dNTP mix (2 m*M* dATP, dGTP, dCTP and 0.2 m*M* dUTP).
   d. *Taq* DNA polymerase (*see* **Note 3**).
   e. Primers and template for amplification of an arbitrary target (approx 500 bp).
3. Exonuclease I (Amersham Life Sciences) (*see* **Note 4**).
4. TE pH 7.5: 10 m*M* Tris-HCl, pH 7.5, 1 m*M* EDTA,
5. 4 *M* ammonium acetate, 100% ethanol and 70% ethanol.
6. Centri-spin-10 columns (Princeton Separations Adelphia, NJ) (*see* **Note 5**).
7. Spectrophotometer.
8. Liquid scintillation cocktail that is miscible with aqueous samples (Ecoscint, National Diagnostics, Atlanta, GA).

2.2. Uracil-Containing Plasmid DNA Preparation

1. *E. coli* (*dut ung*) strain (*see* **Note 6**).
2. Circular plasmid DNA. (*see* **Note 7**).
3. QIAGEN Plasmid Kit (QIAGEN Inc., Valencia, CA).
4. TE, pH 7.5.
5. Spectrophotometer.
6. Agarose gel electrophoresis equipment.

2.3. Stock Solutions for UDG Enzyme Assay

1. 10X UDG reaction buffer: 200 m*M* Tris-HCl, pH 8.0, 10 m*M* EDTA, 10 m*M* dithiothreitol (DTT), supplemented with 1 mg/mL bovine serum albumin (BSA).
2. UDG dilution buffer (1X) : 20 m*M* Tris-HCl, pH 8.0, 50 m*M* NaCl, 1 m*M* EDTA, 1 m*M* DTT, 0.1 mg/mL BSA, 50% (v/v) glycerol.
3. 0.8 *M* trichloroacetic acid (TCA). Store at 4°C.
4. Calf thymus DNA, 1 mg/mL (*see* **Note 8**).
5. Liquid scintillation cocktail.
6. 1 *M* NaOH.
7. 1 *M* Tris base.

3. Methods

3.1. Preparation of Tritium-Labeled DNA Substrate

1. Carry out ten 25 µL PCR reactions containing ^{3}H-dUTP at a final concentration of 2 µM per reaction tube (*see* **Note 9**).
2. Combine all 10 PCR reactions into a 1.5-mL microcentrifuge tube.

3. Add an equal volume of 4 *M* ammonium acetate.
4. Add 2.5 vol of ice-cold 100% ethanol.
5. Vortex briefly and incubate at –70°C for 1 h or overnight at –20°C.
6. Precipitated DNA is pelleted by centrifugation at 4°C (11,600*g* × 30 min).
7. Supernatant is discarded.
8. Add 1 mL of ice-cold 70% ethanol to wash pellet.
9. Mix well and repeat **steps 7** and **8**.
10. Remove all traces of liquid by incubating at 55°C for 30 min.
11. Dissolve the DNA pellet in approx 50 µL of TE.
12. Add 1 U of Exonuclease 1 and incubate at 37°C for 30 min followed by 80°C for 15 min to inactivate the enzyme.
13. Remove any unincorporated dNTPs via column purification using Centri-Spin-10 columns .
14. Measure the absorbance at 260 nm to determine the DNA concentration.
15. Determine the specific radioactivity (cpm/µg DNA, cpm/pmol DNA, or cpm/pmol uracil) by counting a sample of the DNA in the tritium channel of a scintillation counter.

3.2. UDG Assay on Tritium-Labeled PCR Template

The assay measures the release of [^{3}H] - uracil from the PCR substrate. [^{3}H] uracil released by UDG is acid soluble in cold TCA, wheras the resulting abasic DNA, being insoluble, precipitates and is removed via centrifugation.

1. Set up two 1.5-mL microcentrifuge tubes containing the following;
   a. 5 µL 10X UDG reaction buffer.
   b. ^{3}H-labeled DNA (approx 1000 cpm).
   c. Sterile water to 50 µL.
2. Incubate at 37°C for 15 min.
3. Initiate the reaction by adding varying amounts of UDG enzyme (prepared in dilution buffer) to tube 1.
4. Add an equivalent amount of dilution buffer only to tube 2 (*see* **Note 10**).
5. Incubate at 37°C for 15 min.
6. Add 5 µL of calf thymus DNA (1 mg/mL).
7. Stop the reaction by adding an equal volume of ice-cold 0.8 *M* TCA.
8. Vortex briefly and place on ice for 15 min to allow the DNA to precipitate.
9. Spin at 11,600*g* for 15 min in a microcentrifuge at 4°C.
10. Remove the supernatant (being careful to ensure the precipitated DNA is undisturbed) to a scintillation vial and add 3 mL of scintillation fluid.
11. Mix well and count in the ^{3}H-channel of the scintillation counter.
12. Subtract the background cpm (tube 2) from the cpm for tube 1 and express UDG activity as pmols uracil released per minute from dsDNA per microgram protein (*see* **Note 11**).

3.3. UDG Assay Using Plasmid DNA Substrates Containing dUMP Residues

The *E. coli* (*dut ung*) strain is transformed with plasmid DNA using standardprocedures. Antibiotic resistant clones are selected and plasmid DNA purified using one of a number of commercially available products, e.g., QIAGEN Kit or using standard laboratory protocols. The DNA concentration is determined by absorbance measurement at 260 nm. This DNA is used as substrate for UDG activity as described below.

The assay exploits the fact that dUMP containing plasmid DNA is susceptible to glycolytic cleavage at uracil bases by UDG and the phosphate backbone at the resulting abasic sites are cleaved when incubated under alkali conditions at 95°C.

1. Preincubate approx 100 ng plasmid DNA, in 1X UDG reaction buffer at 37°C for 10 min. (*see* **Note 12**).
2. Initiate the reaction by adding varying amounts of UDG, (purified protein, protein fraction, or cell extract).
3. Mix and place at 37°C for 30 min.
4. Add NaOH to a final concentration of 50 m*M* and incubate at 95°C for 15 min.
5. Add Tris base to a final concentration of 30 m*M* (*see* **Note 13**).
6. Separate digestion products on an agarose gel and visualize the DNA by ethidium bromide staining (*see* **Note 14**).

4. Notes

1. Both circular and linear forms of DNA are susceptible to digestion by UDG. Therefore, plasmid linearization prior to digestion with UDG is not necessary.
2. The [^{3}H] nucleotide is supplied as an ethanol:water (1:1) preparation. The ethanol can be removed by applying a gentle stream of N_2 to the surface of the solution.
3. To ensure adequate incorporation of dUTP, it is recommended that high-fidelity polymerases not be used in the amplification reactions. Proofreading enzymes, which contain a 3'–5' exonuclease activity, do not incorporate dUTP efficiently.
4. The 3'-5' exonuclease activity of Exonuclease 1 acts to eliminate any residual single-stranded DNA remaining after PCR amplification.
5. CentriSpin-10 purification columns serve to remove unincorporated nucleotides and nucleotides generated through the action of Exonuclease 1.
6. *E. coli* (*dut ung*) double mutants are deficient in dUTPase and UDG activities, therfore dUMP is readily incorporated and maintained during replication in these hosts. For our assays, we used *E. coli* BW313 (λ^-, *relA*1, *ung* $^{-1}$, *dut* $^{-1}$, *spot*1, *thi-)* (*E. coli* Genetic Stock Center, Dept. of Biology, 355 OML, Yale University, New Haven, CT).
7. Any double-stranded plasmid DNA may be used as substrate.
8. The inclusion of calf thymus DNA in the precipitation reaction serves to enhance precipitation of small amounts of DNA fragments (<100 ng DNA).

9. Care should be exercised in the handling and disposal of tritium. Wear suitable gloves at all times. Dispose of any liquid waste or solid waste as radioactive waste.
10. Tube 2 serves as a control tube containing no enzyme and it normalizes for any acid-soluble [^{3}H]-dUTP that failed to be removed during the template purification procedure. The total cpms observed here should be extremely low and subtracted from the total cpms observed in all reactions.
11. Determine the DNA concentration using:
    a. $Abs_{260nm} \times 50 \times$ dilution factor = µg/mL.
    b. Convert to molar nucleotide residues: µg/mL × 330 × total number of residues.
    c. Count the radioactivity in a small amount of [^{3}H] DNA (approx 5 µL).
    d. Express as cpm/pmol [^{3}H] DNA.

    In terms of dUMP residues: cpm/pmol dUMP residues = pmol DNA × number of dUMP residues per DNA molecule.

    Define 1 unit of UDG activity as cpm equivalent to the release of Xpmol uracil/min/µg protein assayed.
12. a. To ensure there is no contaminating nuclease activity, a reaction containing control plasmid DNA (i.e., containing dTMP instead of dUMP residues) should be set up. This is especially important when one is assaying for UDG activity in more impure enzyme preparations. Any digestion observed here indicates the presence of nonspecific nucleases.
    b. A control reaction should be set up where no UDG or protein is added, but the NaOH step (**step 4**) is carried out. This allows any background nicking of the DNA in the absence of glycolyase activity to be taken into account.
13. The Tris base serves to lower the pH because the alkali can alter the electrophoresis of DNA.
14. An undigested control plasmid DNA sample should be electrophoresed alongside digested samples. Uracil containing plasmid substrates, digested with UDG, and posttreated with NaOH at 95°C, migrate as DNA smears through agarose gels. Following use of higher UDG concentrations, smaller DNA fragments are produced that migrate faster as diffuse bands and are well separated from undigested DNA.

References

1. Sancar, A. (1996) DNA excision repair. *Annu. Rev. Biochem.* **65,** 43–81.
2. Lindahl, T. (1974) An N-glycosidase from *Escherichia coli* that releases free uracil from DNA containing deaminated cytosine residues. *Proc. Natl. Acad. Sci. USA* **71,** 3649–3653.
3. Lindahl, T., Ljingquist, S., Siegert, W., et al. (1977) DNA N-glycosidases: properties of uracil-DNA glycosidase from *Escherichia coli. J. Biol. Chem.* **252,** 3286–3294.
4. Bennett, S. E. and Mosbaugh, D. W. (1992) Characterisation of the *Escherichia coli* uracil DNA glycosylase inhibitor protein complex. *J. Biol. Chem.* **267,** 22,512–22,521.
5. Karran, P., Cone, R., and Friedberg, E. C. (1981) Specificity of the bacteriophage PBS2 induced inhibitor of uracil-DNA glycosylase. *Biochemistry* **21,** 6092–6096.

4

Detection of DNA Deoxyribophosphodiesterase Activity

Margarita Sandigursky and William A. Franklin

1. Introduction

The major pathway for the removal of oxidative base damage is the DNA base excision repair pathway, found in prokaryotes and eukaryotes *(1)*. In this pathway, oxidized DNA bases are removed by specific DNA glycosylases, leaving apurinic/apyrimidinic (AP) sites in the DNA *(1,2)*. AP sites can also arise spontaneously in DNA through depurination (*3*) and, being devoid of genetic information, can be both cytotoxic and mutagenic lesions *(4–6)*. Several DNA glycosylases have been found that convert a variety of damaged nucleotide residues to AP sites by removing deaminated, oxidized, or alkylated bases from DNA. Uracil, either misincorporated in place of thymine or resulting from deamination of cytosine, is removed by a specific glycosylase, uracil-DNA glycosylase, found in prokaryotes and eukaryotes *(1,4)*.

AP sites are recognized by enzymes that cleave the site either on the 5' side or the 3' side. Enzymes that incise the DNA at the 5' side of an AP site, AP endonucleases, leave 5' 2-deoxyribose-5-phosphate termini (5' dRp) and 3' OH groups. In *Escherichia coli*, the reaction is catalyzed by the AP endonuclease activity of exonuclease III, which accounts for nearly 90% of the activity detected in the bacteria *(2,7,8)*, or by the inducible endonuclease IV *(2,9)*. The major AP endonucleases isolated from mammalian cells seem to act in a similar manner *(2)*. Enzymes that incise the DNA at 3' side of an AP site, AP lyases, cleave the DNA by a β-elimination reaction, and leave *trans*-4-hydroxy-2-pentenal-5-phosphate termini and 5' phosphate groups *(2,10)* (*see* **Note 1**). AP lyases are activities found associated with several glycosylases such as endonuclease III of *E. coli* and its human analog *(11–13)*, the *E. coli* Fpg protein *(14)*, and the yeast and human Ogg1 proteins *(15,16)*.

From: *Methods in Molecular Biology, vol. 152: DNA Repair Protocols: Prokaryotic Systems*
Edited by: P. Vaughan © Humana Press Inc., Totowa, NJ

Removal of either the *trans*-4-hydroxy-2-pentenal-5-phosphate group following incision by an AP lyase, or the 5' dRp group following incision by an AP endonuclease is required to allow for the subsequent replacement of the AP site by one or more nucleotides by the activity of a DNA polymerase. Enzymes that specifically remove the sugar-phosphate groups left by AP endonucleases/lyases are termed DNA deoxyribophosphodiesterases, or dRpases, and were first described in *E. coli* ***(17)***. Although dRpase was first used to describe hydrolytic activities that remove 5' dRp residues at incised AP sites ***(17)***, the term has also been used to describe phosphodiesterase activities that remove *trans*-4-hydroxy-2-pentenal-5-phosphate groups, as well as activities that remove the 5' dRp residues via nonhydrolytic catalysis, for example, DNA polymerase β ***(18)***. In *E. coli*, dRpase activities that remove 5' dRp groups include exonuclease I ***(19)***, RecJ ***(20)***, and the Fpg protein ***(21)***. Eukaryotic activities include a partially purified activity from HeLa cells ***(22)***, an activity associated with DNA polymerase β ***(18,23)***, the *Drosophila* S3 protein ***(24)*** and the yeast Ogg1 ***(25)***. Activities in *E. coli* that remove *trans*-4-hydroxy-2-pentenal-5-phosphate groups include exonuclease I ***(19)***, exonuclease III and endonuclease IV ***(26,27)***, and exonuclease IX ***(28)***. Eukaryotic activities include the yeast Apn1 protein ***(29)***, the yeast Ogg1 protein ***(25)***, and the *Drosophila* S3 protein ***(24)***.

To assay dRpase activity, double-stranded DNA substrates are employed that contain radiolabeled incised AP sites. Radiolabeled dUMP residues are incorporated into the DNA, and DNA is subsequently treated with uracil-DNA glycosylase and either an AP endonuclease (*E. coli* endonuclease IV) or an AP lyase (*E. coli* endonuclease III). Release of the radiolabeled sugar-phosphate groups is either measured indirectly in a rapid precipitation assay or directly by chromatographic separation on anion-exchange high-performance liquid chromatography (HPLC). One problem with using radiolabeled susbstrates is that the incorporated radioactivity itself can cause DNA strand breaks and can also induce the loss of the sugar-phosphate groups. To diminish this effect, we now use the low-energy β-emitter [^{33}P] (0.25 MeV) in place of the more prevalently used [^{32}P] (1.7 MeV).

2. Materials

2.1. Reagents

All reagents were from Sigma-Aldrich, except where noted.

1. Deoxycytidine 5'-triphosphate [α-^{33}P], 2000–4000 Ci/mmol, aqueous solution buffered with Tricene, pH 7.6, dye free, from New England Nuclear Life Science Products, Boston, MA.
2. Phosphorus-33 radionuclide, 40–158 Ci/mmol, from New England Nuclear Life Science Products.

3. 1 *M* nitrous acid (prepare fresh before use): 69 mg $NaNO_2$ in 1 mL 1 *M* sodium acetate buffer, pH 4.0.
4. Norit activated carbon, Norit SA-3–100 mesh, decolorizing.
5. [α-^{33}P] dUTP elution buffer: 3.5% NH_3 in 50% ethanol (14.5 mL 25% amonium hydroxide, 35.5 mL water, 50 mL ethanol).
6. 0.45 μm Versapack LC3A filter, Gilman Sciences, Ann Arbor, MI.
7. Single-stranded M13mp18, 0.2 μg/μL, Pharmacia Biotech.
8. M13 sequencing primer (-47), 24-mer: 0.5 μ*M* (3 ng/μL) in water.
9. 10X Klenow buffer: 0.5 *M* Tris-HCl, pH 7.6, 0.1 *M* $MgCl_2$.
10. Deoxynucleotides 5'-triphosphates: 1 m*M* dATP, 1 m*M* dCTP, 1 m*M* dGTP, 1 m*M* dTTP, 0.1 m*M* dTTP, 0.1 m*M* dUTP, from Boehringer Mannheim.
11. Quick Spin (TE) Columns [Sephadex G-50 (DNA)] from Boehringer Mannheim.
12. 10X uracil-DNA glycosylase buffer: 0.5 *M* HEPES-NaOH, pH 7.8, 10 m*M* Na_2 ethylenediaminetetraacetic acid (EDTA), 5 m*M* DTT.
13. 3 *M* sodium acetate, pH 5.2.
14. 95% ethanol.
15. 10 m*M* HEPES-NaOH, pH 7.8
16. 10X DNA dRpase buffer: 0.5 *M* HEPES-NaOH, pH 6.6, 0.1 *M* $MgCl_2$, 5 m*M* dithiothreitol (DTT).
17. 10% (w/v) solution of trichloroacetic acid (TCA).
18. Calf thymus DNA: 2.5 mg/mL in 10 m*M* Tris-HCl, pH 7.8, 1 m*M* Na_2 EDTA.
19. 5% (w/v) Norit activated carbon in water.
20. HPLC elution buffer: 25 m*M* KH_2PO_4, pH 3.5 (Fisher Scientific, HPLC grade).
21. 2-deoxyribose-5-phosphate (30 mg/mL in water).
22. 25 m*M* 2,4-dinitrophenylhydrazine (2,4-DNPH): 50 mg 2,4-DNPH dissolved in 100 mL 1 *M* HCl (stable at room temperature).
23. 1 *M* NaOH.
24. 0.5 *M* Na_2 EDTA, pH 8.0.

For all reactions, 1.7-mL Slick (siliconized) microtubes from PGC Scientifics, Gaithersburg, MD, or equivalent were utilized.

2.2. Enzymes

All enzymes are suspensions in 50% glycerol and are stored at –20°C. Enzymes can also be obtained from other suppliers or laboratories.

1. Klenow enzyme, labeling grade (DNA polymerase I, large fragment) from Boehringer Mannheim.
2. *E. coli* uracil-DNA glycosylase (HK-UNG Thermolabile N-Glycosylase), from Epicentre Technologies, Madison, WI.
3. *E. coli* endonuclease III (Pyr Ox cutter) from Trevigen, Inc., Gaithersburg, MD.
4. *E. coli* endonuclease IV (AP cutter II) from Trevigen, Inc.
5. *E. coli* exonuclease I, from Amersham Life Science Products, Arlington Heights, IL.

2.3. Apparatus

1. Liquid scintillation counter—capable of detecting ^{33}P radioactivity, such as a Tri-Carb 1600CA counter (Packard Instruments).
2. HPLC equipment—consisting of one or more pumps, valve injector, fraction collector, and UV light absorbance monitor.
3. Anion-exchange HPLC columns: Brownlee MPLC AX or amino (NH_2) 5-µm (4.6 mm × 3 cm) spheri-5 cartridge column fitted with an amino NewGuard column (7 µm), Alltech Associates, Deerfield, IL.
4. Standard benchtop centrifuges, microcentrifuges, and lyophilizer.

3. Methods

3.1. General Protocol for dRpase Substrates

To prepare substrates to assay dRpase activity, first [^{33}P] dUTP, which is not commercially available, is prepared by deamination of [^{33}P] dCTP with nitrous acid. This radiolabeled nucleoside triphosphate is then incorporated into DNA by the action of the Klenow fragment of *E. coli* DNA polymerase I on single-stranded M13mp18 DNA to which a 24 mer primer has been annealed (***21***). The reaction concentrations of [^{33}P] dUTP and dTTP are controlled to incorporate approximately one dUMP residue per 50–100 basepairs. In this system, only one strand of the resulting double-stranded DNA will contain dUMP residues. Prior to the use of M13 DNA, substrates were typically prepared using poly(dA-dT), in which dUMP residues would be incorporated in both strands (***17,19,22***). The M13-based substrates have been found to be more stable than the poly(dA-dT)-based substrates; they undergo much less degradation following the removal of the uracil by uracil-DNA glycosylase and the cleavage of the AP sites by an AP endonuclease or lyase.

The [^{33}P] dUMP-containing M13 DNA is treated first with *E. coli* uracil-DNA glycosylase to remove the uracil base. The resulting substrate containing AP sites is treated either with *E. coli* endonuclease IV to produce a substrate containing 5' incised AP sites or with *E. coli* endonuclease III to produce a substrate containing 3' incised AP sites. Following incubation of either of these substrates with an enzyme containing dRpase activity will result in the removal of a [^{33}P]-labeled sugar-phosphate product, as shown in **Fig 1**. These products are then separated from the DNA substrate either in a rapid precipitation assay using TCA and Norit activated carbon, or by separation of the entire reaction mixture on anion exchange HPLC. The amount of product released from the substrate is usually expressed as the percent of the product released from the substrate.

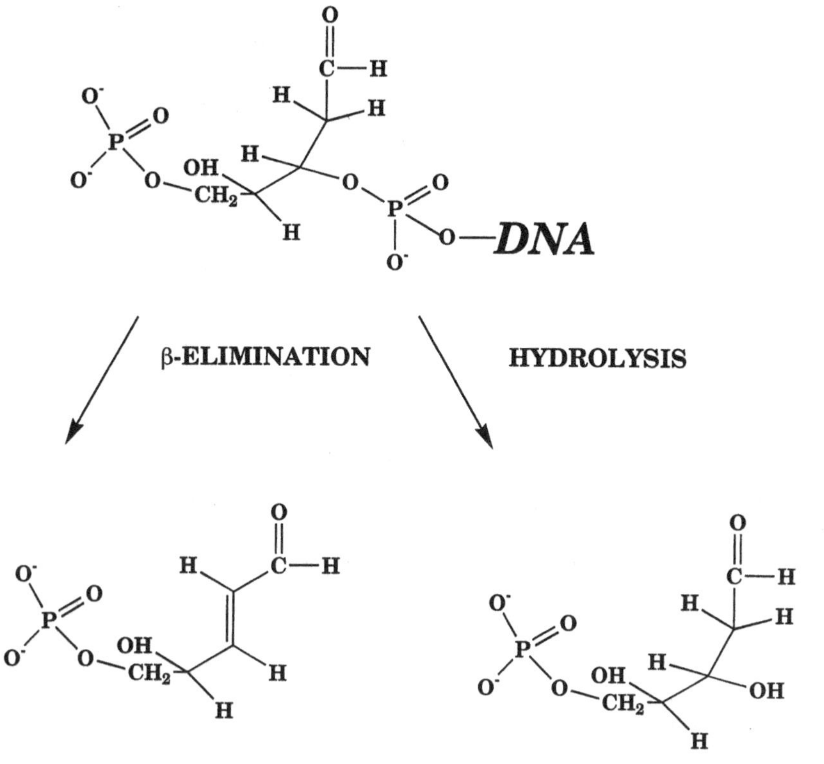

***trans*-4-hydroxy-2-pentenal-5-phosphate** **2-deoxyribose-5-phosphate**

Fig. 1. A dRpase activity catalyzes the release of either the product 2-deoxyribose-5-phosphate by hydrolysis or the product *trans*-4-hydroxy-2-pentenal-5-phosphate by β-elimination from a DNA substrate containing 5' incised AP sites. The product *trans*-4-hydroxy-2-pentenal-5-phosphate is released from a DNA substrate containing 3' incised AP sites.

3.2. Preparation of Radiolabeled [^{33}P] dUTP

1. Incubate 200 μCi (20 μL) of [α-^{33}P] dCTP with 1 mL 1 *M* nitrous acid at 37°C overnight (*see* **Note 2**).
2. Transfer the reaction mixture to a new tube containing 5–6 mg of Norit-activated carbon and vortex for 10–20 s.
3. Centrifuge at 16,000*g* in a microcentrifuge for 5 min at room temperature, then remove and discard the supernatant as radioactive waste.
4. Wash the pellet as follows: resuspend the pellet in 1 mL water, vortex for 5–10 s, and centrifuge for 5 min at 16,000*g* at room temperature. Remove and discard the supernatant as radioactive waste. Repeat this procedure three additional times. The deaminated dCTP ([^{33}P] dUTP) is adsorbed to the carbon.

5. Add 1 mL of the [α-^{33}P] dUTP elution buffer to the pellet, vortex for 5–10 s, and allow to stand at room temperature for 20 min with occasional vortexing.
6. Centrifuge at 16,000*g* in a microcentrifuge at room temperature. Transfer the solution to a 1-mL syringe to which is attached the 0.45-μm Versapack LC3A filter, and pass the solution through the filter.
7. Lyophilyze the [α-^{33}P] dUTP solution to dryness.
8. Dissolve the lyophilyzed [α-^{33}P] dUTP product in 200 μL of 2 m*M* phosphate buffer, pH 7.0. Measure the radioactivity of a 1-μL solution. Approximately 30% of the initial radioactivity is recovered as [α-^{33}P] dUTP.
9. The [α-^{33}P] dUTP solution is stored at –20°C.

3.3. Preparation of M13 DNA Substrates Containing [^{33}P] dUMP Residues

1. Determine the radioactivity of 1 μL of the [α-^{33}P] dUTP solution. Determine the volume containing 15 μCi [α-^{33}P] dUTP (2.22 × 10^{12} dpm/Ci).
2. Combine 15 μCi [α-^{33}P] dUTP, 5 μL (1 μg) M13mp18 DNA , 10 μL M13 24-mer primer (30 ng), 2.5 μL 1 m*M* dATP, 2.5 μL 1 m*M* dCTP, 2.5 μL 1 m*M* dGTP, 2.5 μL 0.1 m*M* dTTP, 1.5 μL 0.1 m*M* dUTP, 10 μL 10X Klenow buffer, 3 U Klenow enzyme and water to a total reaction volume of 100 μL. Incubate the mixture for 2 h at 37°C.
3. Supplement the reaction mixture with an additional 2.5 μL 1 m*M* dATP, 2.5 μL 1 m*M* dCTP, 2.5 μL 1 m*M* dGTP, 2.5 μL 1 m*M* dTTP, and 2 U of Klenow enzyme. Incubate for an additional 1 h at 37°C
4. Stop the reaction by the addition of 5 μL 0.5 *M* Na_2 (EDTA). Heat the reaction mixture for 10 min at 70°C. Chill on ice.
5. Unincorporated nucleotides are removed from the [^{33}P] dUMP-containing M13 DNA by the use of a Quick-Spin (Sephadex G-50) column. Allow the buffer in the prepacked column to drain by gravity, then centrifuge the column in a swinging bucket rotor at 1100*g* for 2 min at room temperature. Discard the eluted buffer. Apply the reaction mixture (up to 100 μL) to the center of the column bed. Centrifuge the column in a swinging bucket rotor at 1100 *g* for 4 min at room temperature. Measure the volume of the eluate; discard the column as radioactive waste.
6. Precipitate the [^{33}P] dUMP-containing M13 DNA by the addition of 0.1 vol of 3 *M* acetate buffer, pH 5.2, and 2.2 vol of 95% ethanol. Chill at –20°C for 20 min and centrifuge at 16,000*g* in a microcentrifuge at 4°C for 15 min. Discard the supernatant and add 900 μL 70% ethanol. Centrifuge at 16,000*g* in a microcentrifuge at room temperature for 1 min and discard the supernatant.
7. Lyophilize the DNA to dryness.
8. Dissolve the [^{33}P] dUMP-containing M13 DNA in 10 μL 10X uracil-DNA glycosylase buffer, 90 μL water.

3.4. Preparation of M13 DNA Substrates with 5' Incised AP Sites

1. Add 0.5 U *E. coli* uracil-DNA glycosylase to 100 μL [^{33}P] dUMP-containing M13 DNA. Incubate the mixture at 37°C for 40 min.

2. Add 1 U of *E. coli* endonuclease IV to the reaction mixture and incubate for an additional 40 min.
3. Stop the reaction by heating the mixture at 70°C for 15 min. Chill on ice.
4. Precipitate the DNA by the addition of 0.1 vol of 3 *M* acetate buffer, pH 5.2, and 2.2 vol of 95% ethanol. Chill at –20°C for 20 min and centrifuge at 16,000*g* in a microcentrifuge at 4°C for 15 min. Discard the supernatant and add 900 µL 70% ethanol. Centrifuge at 16,000*g* in a microcentrifuge at room temperature for 1 min and discard the supernatant.
5. Lyophilize the DNA to dryness.
6. Dissolve the DNA in 200 µL 10 m*M* HEPES-NaOH, pH 7.6. Measure the radioactivity of 1 µL of the sample.
7. Store the substrate at –20°C (*see* **Note 3**).

3.5. Preparation of M13 DNA Substrates with 3' Incised AP Sites

Follow the procedure in **Subheading 3.4.**, except substitute 1 U of *E. coli* endonuclease III in place of 1 U of *E. coli* endonuclease IV in **step 2** (*see* **Note 3**).

3.6. DNA dRpase Activity Assays to Detect the Release of Sugar-Phosphate Products

3.6.1. Precipitation Assay

1. Combine 15,000–20,000 dpm M13 DNA containing incised AP sites, 10 µL 10X DNA dRpase buffer, 1 U *E. coli* exonuclease I (or another enzyme with dRpase activity), and water to a total reaction volume of 100 µL. Incubate the mixture for 30 min at 37°C. A control reaction should be performed in the absence of enzyme.
2. Add 11 µL of ice-cold calf thymus DNA, 110 µL 10 % TCA solution, and 30 µL 5% Norit-activated carbon and vortex for 5–10 s.
3. Centrifuge at 16,000*g* for 10 min in a microcentrifuge at 4°C.
4. Carefully remove the supernatant from the pellet. Transfer the supernatant to a liquid scintillation vial and determine the amount of radioactivity in the sample. This value divided by the amount of the total radioactivity contained in the substrate will give the fraction of sugar-phosphate product released by the enzyme.

3.6.2. HPLC Assay

3.6.2.1. Equilibration of the Anion-Exchange Column and Determination of the Elution Positions of Sugar-Phosphate and Phosphate

1. Equilibrate the Brownlee MPLC AX or amino (NH_2) spheri-5 cartridge column fitted with an amino NewGuard column with at least 25 mL of the HPLC elution buffer. Monitor the elution of buffer with a UV detector at 260 nm until a stable baseline is reached and the column maintains a stable back pressure.

2. To determine the relative elution positions of sugar-phosphate and phosphate from the column, mix 90 μL of 2-deoxyribose-5-phosphate and 10 μL phosphorus 33 containing 10,000–15,000 cpm. Inject this mixture onto the column and elute with the HPLC elution buffer at a flow rate of 1 mL/min. Collect fractions every 0.5 min. Collect at least 40 fractions (20-mL elution).
3. Count the radioactivity contained in each fraction to determine the elution position of inorganic phosphate.
4. To determine the elution position of sugar-phosphate, transfer an aliquot of 150 μL from each fraction to a new tube. To each tube, add 100 μL of 25 m*M* 2,4-DNPH and allow to stand for 5 min at room temperature. Add 500 μL of 1 *M* NaOH to each tube. Samples containing sugar-phosphate (2-deoxyribose-5-phosphate) will turn color from yellow to brick red.

3.6.2.2 HPLC dRpase Activity Assay

1. Follow **step 1** in **Subheading 3.6.1.**
2. Inject the 100-μL reaction mixture on the HPLC column and elute with the HPLC elution buffer at a flow rate of 1 mL/min. Collect 40 fractions every 0.5 min.
3. Determine the radioactivity contained in each fraction.

4. Notes

1. Both 2-deoxyribose-5-phosphate and *trans*-4-hydroxy-2-pentenal-5-phosphate coelute from the anion-exchange columns. If a M13 DNA substrate containing 3' incised AP sites is used in the dRpase assay, the released product will always be *trans*-4-hydroxy-2-pentenal-5-phosphate (unless there is contamination with a phosphatase). However, release of the sugar-phosphate product from the M13 DNA substrate containing 5' incised AP sites can be 2-deoxyribose-5-phosphate if the enzyme releases the product by hydrolysis or *trans* -4-hydroxy-2-pentenal-5-phosphate if the enzyme releases the product via a β-elimination mechanism. If the enzymatic mechanism of the release of the sugar-phosphate from the substrate containing 5' incised AP sites is unknown, the chemical product can be determined by including sodium thioglycolate in the reaction mixture. Sodium thioglycolate will form compounds with the sugar-phosphate products released during a β-elimination reaction. These products can then be analyzed by anion-exchange HPLC (***21,24***).
2. It is certainly possible to substitute [α-^{32}P] dCTP in place of [α-^{33}P] dCTP. As mentioned in the introduction, the energy of the β electron emitted from [^{32}P] is approx 7 times higher than that of [^{33}P], and incorporation of [^{32}P] will cause increased rates of substrate degradation. The half-life of [^{33}P] is 25 d, as compared to 14 d for [^{32}P]. Furthermore, much more stringent precautions for radiation safety have to be employed when using [^{32}P]. However, [^{32}P]-containing nucleotides are more readily available than [^{33}P]-containing nucleotides, and are considerably less expensive to purchase.
3. The M13 DNA substrate containing 3' incised AP sites is chemically stable. The M13 DNA substrate containing 5' incised AP sites will undergo loss of the

sugar-phosphate product at a rate of 5–10%/d. This may occur from non-enzymatic β-elimination of the sugar-phosphate induced by the presence of the [^{33}P] radionuclide.

References

1. Lindahl, T. (1993) Instability and decay of the primary structure of DNA. *Nature* **362,** 709–715.
2. Doetsch, P. W. and Cunningham, R. P. (1990) The enzymology of apurinic/apyrimidinic endonucleases. *Mutat. Res.* **236,** 173–201.
3. Lindahl, T. and Nyberg, B. (1972) Rate of depurination of native deoxyribonucleic acid. *Biochemistry* **11,** 3610–3618.
4. Olsen, L. C., Aasland, R., Krokan H. E., and Helland D. E. (1991) Human uracil-DNA glycosylase complements *E. coli ung* mutants. *Nucleic Acids Res.* **19,** 4473–4478.
5. Loeb, L. A. (1985) Apurinic sites as mutagenic intermediates. *Cell* **40,** 483–484.
6. Loeb, L. A. and Preston, B. D. (1986) Mutagenesis by apurinic/apyrimidinic sites. *Annu. Rev. Genet.* **20,** 201–230.
7. Yajko, D. M. and Weiss, B. (1975) Mutations simultaneously affecting endonuclease II and exonuclease III in *Escherichia coli. Proc. Natl. Acad. Sci. USA* **72,** 688–692.
8. Weiss, B. (1976) Endonuclease II of *Escherichia coli* is exonuclease III. *J. Biol. Chem.* **251,** 1896–1901.
9. Chan, E. and Weiss, B. (1987) Endonuclease IV of *Escherichia coli* is induced by paraquat. *Proc. Natl. Acad. Sci. USA* **84,** 3189–3193.
10. Mazumder, A., Gerlt, J. A., Absalon, M. J., et al. (1991) Stereochemical studies of the β-elimination reactions at aldehyic abasic sites in DNA: endonuclease III from *Escherichia coli*, sodium hydroxide, and ly-trp-lys. *Biochemistry* **30,** 1119–1126.
11. Bailly, V. and Verly, W. G. (1987) *Escherichia coli* endonuclease III is not an endonuclease but a β-elimination catalyst. *Biochem. J.* **242,** 565–572.
12. Aspinwall, R., Rothwell, D. G., Roldan-Arjona, T., et al. (1997) Cloning and characterization of a functional human homolog of *Escherichia coli* endonuclease III. *Proc. Natl. Acad. Sci. USA* **94,** 109–114.
13. Hilbert, T. P., Chaung, W., Boorstein, R. J., et al. (1997) Cloning and expression of the cDNA encoding the human homologue of the DNA repair enzyme, *Escherichia coli* endonuclease III. *J. Biol. Chem.* **272,** 6733–6740.
14. Boiteux, S., O'Connor, T. R., Lederer, F., et al. (1990) Homogeneous *Escherichia coli* FPG protein. A DNA glycosylase which excises imidazole ring-opened purines and nicks DNA at apurinic/apyrimidinic sites. *J. Biol. Chem.* **265,** 3916–3922.
15. van der Kemp, P. A., Thomas, D., Barbey, R., et al. (1996) Cloning and expression in *Escherichia coli* of the OGG1 gene of *Saccharomyces cerevisiae*, which codes for a DNA glycosylase that excises 7,8-dihydro-8-oxoguanine and 2,6-diamino-4-hydroxy-5-N methylformamidopyrimidine. *Proc. Natl. Acad. Sci. USA* **93,** 5197–5202.

16. Lu, R., Nash, H. M., and Verdine, G. L. (1997) A mammalian DNA repair enzyme that excises oxidatively damaged guanines maps to a locus frequently lost in lung cancer. *Curr. Biol.* **7,** 397–407.
17. Franklin, W. A. and Lindahl, T. (1988) DNA deoxyribophosphodiesterase. *EMBO J.* **7,** 3616–3622.
18. Matsumoto, Y. and Kim, K. (1995) Excision of deoxyribose phosphate residues by DNA polymerase β during DNA repair. *Science* **269,** 699–702.
19. Sandigursky, M. and Franklin, W. A. (1992) DNA deoxyribophosphodiesterase of *Escherichia coli* is associated with exonuclease I. *Nucleic Acids Res.* **20,** 4699–4703.
20. Dianov, G., Sedgwick, B., Daly, G., et al. (1994) Release of 5'-terminal deoxyribose-phosphate residues from incised abasic sites in DNA by the *Escherichia coli* RecJ protein. *Nucleic Acids Res.* **22,** 993–998.
21. Graves, R. J., Felzenszwalb, I., Laval, J., and O'Connor, T. R. (1992) Excision of 5'-terminal deoxyribose phosphate from damaged DNA is catalyzed by the Fpg protein of *Escherichia coli. J. Biol. Chem.* **267,** 14,429–14,435.
22. Price, A. and Lindahl, T. (1991) Enzymatic release of 5'-terminal deoxyribose phosphate residues from damaged DNA in human cells. *Biochemistry* **30,** 8631–8637.
23. Feng, J., Crasto, C. J., and Matsumoto, Y. (1998) Deoxyribose phosphate excision by the N-terminal domain of the polymerase β: the mechanism revisited. *Biochemistry* **37,** 9605–9611.
24. Sandigursky, M., Yacoub, Y., Kelley, et al. (1997) The Drosophila ribosomal protein S3 contains a DNA deoxyribophosphodieserase (dRpase) activity. *J. Biol. Chem. 272,* 17,480–17,484.
25. Sandigursky, M., Yacoub, A., Kelley, M. R., et al. (1997) The yeast 8-oxoguanine DNA glycosylase (Ogg1) contains a DNA deoxyribophosphodiesterase (dRpase) activity. *Nucleic Acids Res.* **25,** 4557–4561.
26. Demple, B., Johnson, A., and Fung, D. (1986) Exonuclease III and endonuclease IV remove 3' blocks from DNA synthesis primers in H_2O_2-damaged *Escherichia coli. Proc. Natl. Acad. Sci. USA* **83,** 7731–7735.
27. Bernelot-Moens, C. and Demple, B. (1989) Multiple DNA repair activities for 3'-deoxyribose fragments in *Escherichia coli. Nucleic Acids Res.* **17,** 587–600.
28. Shafritz, K. M., Sandigursky, M., and Franklin, W. A. (1998) Exonuclease IX of *Escherichia coli. Nucleic Acids Res.* **26,** 2593–2597.
29. Popoff, S. C., Spira, A. I., Johnson, A. W., and Demple, B. (1990) Yeast structural gene (APN1) for the major apurinic endonuclease: homology to *Escherichia coli* endonuclease IV. *Proc. Natl. Acad. Sci. USA* **87,** 4193–4197.

5

O^6-Alkylguanine-DNA Alkyltransferase Assay

Amanda J. Watson and Geoffrey P. Margison

1. Introduction

The biological effects of alkylating agents in both pro- and eukaryotes are thought to be mediated via alkylation at the O^6-position of guanine in DNA *(1–4)*. Repair of such adducts can be mediated by O^6-alkylguanine-DNA alkyltransferase (ATase; *3,4*). Both pro- and eukaryote ATases transfer alkyl groups from the O^6-position of guanine in alkylated DNA (or from other low-molecular-weight substrates) *(5)* to a cysteine residue located at the active site of the protein: the reaction is stoichiometric and the protein is autoinactivated *(6)*. This mechanism has been exploited in the design of several different radioactivity-based assays for the enzyme. These involve either measurement of methyl group transfer to protein or the analysis [e.g. by high-performance liquid chromatography (HPLC)] of methylated substrate DNA before and after exposure to cell or tissue extracts or restriction endonuclease (RE) site deprotection of synthetic oligonucleotide substrates containing O^6-methylguanine.

The most common method of assaying ATase activity in cell or tissue extracts, which we have used successfully and propagated to a number of other laboratories over a number of years (e.g., **ref.** *7*), involves measuring [^{3}H] methyl group transfer to ATase protein. Essentially, high specific radioactivity [^{3}H]-methylated DNA substrate is incubated with extract under protein-limiting conditions until the transfer reaction is complete. Excess substrate DNA is hydrolysed to acid solubility and radioactivity in the residual protein is measured by liquid scintillation counting.

The availability of very high specific activity ^{32}P- and ^{35}S-labeled deoxynucleoside triphosphates has encouraged the development of more sensitive ATase assays based on end-labeled fragments of oligonucleotides con-

From: *Methods in Molecular Biology, vol. 152: DNA Repair Protocols: Prokaryotic Systems*
Edited by: P. Vaughan © Humana Press Inc., Totowa, NJ

taining O^6-alkylguanine sometimes in a RE site. A number of different methods have been devised ***(8–12)***, but a convenient, highly sensitive alternative to the standard methyl transfer method is not, as yet, in widespread use.

2. Materials

2.1. Preparation of Substrate DNA

2.1.1. Deproteinization of DNA

1. Calf thymus DNA (*see* **Note 1**).
2. Duran (or other wide-necked glass) bottles. Because of the hazards associated with this procedure (*see* **Note 2**) minimize the possibility of leakage by ensuring that the bottles have a good seal.
3. Proteinase K.
4. Phenol equilibrated with 1 *M* Tris-HCl, pH 8.0. Prepare fresh as required. Add an equal volume of 1 *M* Tris-HCl, pH 8.0 to the phenol. Shake, allow to settle, and aspirate off as much of aqueous phase as possible. Extreme caution must be exercised when handling and disposing of phenol (*see* **Note 2**).
5. 3 *M* NaAc, pH 4.0.
6. Absolute ethanol.
7. Ether.
8. N_2.
9. Water bath set at 55°C.
10. 5 mL plastic syringe.
11. Water-pump aspirator.
12. 50-mL Falcon tubes (conical bottom).

2.1.2. Methylation of DNA

1. Duran (or other wide-necked glass) bottles (*see* **Subheading 2.1.1.**).
2. 0.02 *M* Ammediol (2-amino-2-methyl-1-3-proponediol), pH 10.0 (Sigma) (pH adjustment is not necessary).
3. 5 mCi [^{3}H] MNU. Preferably approx 20 Ci/mmol (Amersham International, 1 mCi/mL in ethanol). Use immediately on delivery.
4. Absolute ethanol.
5. Pasteur pipets.
6. Water vacuum aspirator.
7. Ether.
8. Ethanol: ether (1:1).
9. N_2.
10. Chemical fume cupboard.

2.1.3. Checking New Substrate

1. Buffer I : 50 m*M* Tris-HCl, pH 8.3 EDTA, (1 m*M* ethylendiaminetetraacetic acid), 3 m*M* (dithiothreitol) (DTT).

This buffer, prepared without DTT, may be stored for several months at 4°C and used for up to 1 wk following the addition of DTT.

2. 5-mL syringes.
3. 21-Gage syringe needles.
4. Ecoscint (National Diagnostics/Mensura).
5. Scintillation counter.

2.2. Preparation of Cell/Tissue-Free Extracts

1. Buffer I (*see* **Subheading 2.1.3.**).
2. PBS (phosphate-buffered saline): 0.8% NaCl, 0.02% KCl, 0.15% $Na_2H_2PO_4$, 0.02% KH_2PO_4, pH 7.4.
3. PMSF (phenylmethylsulphonyl fluoride; Sigma), 50 m*M* in 100% ethanol. Store at –20°C. Stable for at least 3 mo.
4. Leupeptin (Sigma), 10 mg/mL in ddH_2O. Store at 4°C. Stable for at least 1 mo.
5. Sonicator fitted with microtip probe suitable for 1.5-mL Eppendorf tubes (*see* **Note 3**).

2.3. Protein Estimation

1. CBG (Coomassie brilliant blue G250) dye reagent concentrated stock: 780 m*M* CBG (Sigma, 75% dye content), 25% (v/v) ethanol (BDH analar), 7.4 *M* orthophosphoric acid, 0.01% Triton X-100 (v/v), 0.01% SDS (w/v). Store at 4°C in the dark for up to 1 yr.
   Before use, dilute to 1 in 5 in ddH_2O, leave at 4°C overnight then filter through 3-mm filter paper. Store at 4°C in the dark for up to 3 mo. Commercial reagents are available (*see* **Note 4**).
2. IBSA: 1 mg/mL bovine serum albumin (BSA) in buffer I. Store at 4°C.
3. BSA protein standards: standards of 0.1, 0.2, 0.3, 0.4, 0.5, 0.6, 0.8, 1.0 mg/mL BSA in buffer I are prepared from IBSA. Standards are filtered (0.2 microns) and stored for up to 6 mo at 4°C.
4. Scintillation minivials.
5. Plastic spectrophotometer cuvets.
6. Multidispense pipet (*see* **Note 5**).
7. Spectrophotometer set to read at absorbance 595 nm.

2.4. DNA Estimation

1. 10X TNE Buffer : 100 m*M* Tris base, 10 m*M* EDTA, 2 m*M* NaCl, pH 7.4. Filter before use (0.45 μ*M*) and store at 4°C for up to 3 mo.
2. Calf thymus DNA standard (Pharmacia Biotech, ultrapure), 100 μg/mL in TNE. Store at 4°C for up to 3 mo.
3. Hoechst 33258 (*bis*-benzamide) stock dye solution, 1 mg/mL in ddH_2O. Store at 4°C in the dark for up to 6 mo.
4. TKO 100 minifluorometer and fluorometer cuvet (Hoefer, *see* **Note 6**) .

2.5. Assay

1. [^{3}H] methylated calf thymus DNA (*see* **Subheading 2.1.**).
2. IBSA (*see* **Subheading 2.3.**).
3. BSA, 10 mg/mL in ddH_2O. Store at 4°C
4. PCA (perchloric acid), 1 *M* and 4 *M* in ddH_2O.
5. NaOH, 10 m*M* in ddH_2O.
6. Ecoscint (National Diagnostics/Mensura).
7. Scintillation minivials.
8. Dry heat blocks set at 37°C and 75°C.
9. Multidispense pipet (*see* **Note 5**).
10. Scintillation counter.

3. Methods

3.1. Preparation of Substrate DNA

3.1.1. Deproteinization of DNA

1. Dissolve CT DNA at 2 mg/mL in TE (up to 300 mL) on a stirrer overnight in a IL Duran bottle. There will be some insoluble bits, but it is not necessary to remove them.
2. Place bottle in 55°C water bath for 5 min then add solid proteinase K (30 mg). The bits should disappear quickly, but leave for 1 h swirling occasionally before adding another 30 mg of proteinase K.
3. After a further 1 h at 55°C move to fume cupboard on tray, cool under running tap water and add equal volume of phenol equilibrated to pH 8.0 using 1 *M* Tris-HCl (*see* **Subheading 2.1.1.**). Cap and shake vigorously for 5 min—be aware of the possibility of leakage.
4. Allow to stand for about 1.5 h at room temperature: the phenol should settle out and can be almost completely removed by aspiration.
5. Decant supernatant into 50-mL Falcon tubes (conical bottom) and spin at 1000*g*, room temperature, 10 min.
6. Observe interface carefully: if clear, reextraction is not necessary (*see* **Note 7**). Remove phenol from bottom of tube using glass 5-mL pipet or Pasteur pipet and rubber pipet bulb or pipet pump. Do not worry about taking some of the aqueous layer. Pour off supernatants into Duran bottle of appropriate capacity.
7. Add 1/10 vol of 3 *M* NaAc pH 4.0 to pooled aqueous phases, mix well, and add 2 vol cold ethanol. Cap and mix by inversion.
8. Lift out DNA on glass pipet and transfer to smaller Duran. Wash three times with ethanol at room temperature by vigorous shaking and water vacuum pump aspiration of the ethanol. Make sure DNA spreads out in ethanol to ensure complete penetration of ethanol.
9. Wash at room temperature three times with ethanol:ether (1:1) and then three times with ether alone. Each time, pour off the washes into a tray in a fume cupboard (no naked flames/electrical appliances).
10. Dry DNA in stream of N_2 to remove the ether, teasing apart fibrous DNA with Pasteur pipets. Dry to constant weight.

3.1.2. Methylation of DNA

Because of the radiochemical hazard involved, the following procedure should be carried out in a fume cupboard set at maximum flow rate. Handling and disposal of [3H] must be performed in accordance with local rules pertaining to radioactive substances. We advise monitoring for [3H] contamination of the work area before starting and, of course, on completion.

1. For 5 mCi [^{3}H]-MNU in 1 mL ethanol: dissolve 40 mg DNA on a stirrer plate overnight at 8 mg/mL in 0.02 *M* Ammediol, pH 10.0 in a 25 mL Duran bottle. Transfer 2 mL of this solution to a separate container. (This is to be used for rinsing [^{3}H] vial—*see* **step 3**).
2. In tray in fume cupboard, CAREFULLY remove seals from [^{3}H]MNU vial using blunt forceps. Use a twisting rather than pulling action and put aluminium ring and sealing disk directly into beaker in tray. Recap vial with black plastic cap provided—avoid shaking.
3. Put Duran on stirrer in tray, and using 5-mL plastic syringe, carefully transfer MNU solution into stirring DNA solution. Rinse out vial with two 1-mL aliquots of DNA solution by serial transfer. Put syringe and empty vial in sink for careful rinsing.
4. Continue stirring DNA for 5 h at room temperature. Carefully remove stirrer bar then add 1/10 vol 3 *M* NaAc, pH 4.0, and 2 vol of cold ethanol. Form DNA precipitate by swirling and inversion being very careful of leakage—any spills will contain [^{3}H]-methanol, which will blow off rapidly.
5. When DNA has formed a tight ball, carefully remove supernatant using a Pasteur pipet attached to a water vacuum pump the outflow of which is piped directly into the sink drain hole to avoid splashing. Do not be concerned about tiny fragments of DNA being sucked down the sink, but do avoid the bulk of the DNA! It is possible to remove all of the supernatant safely in this way.
6. Wash the DNA with ethanol (about 20–30 mL per wash) making sure pellet is "fluffed" out each time. After 10 washes check [^{3}H] radioactivity counts in 500 µL of wash (+3 mL scintillant) to monitor the washing efficiency. Ideally, the last wash should be close to background, but as small fragments of DNA may be present, two consecutive washes with similar cpm is acceptable.
7. Dry DNA by washing in ethanol:ether (1:1) twice and ether twice. DO NOT aspirate supernatants down the sink! Pour them carefully into a stainless steel tray for evaporation and thorough washing down the sink.
8. Blow off residual ether in a slow stream of nitrogen, teasing apart DNA with Pasteur pipets, if necessary. Transfer to preweighed clean glass vial and dry to constant weight.

3.1.3. Checking New Substrate

1. Weigh about 8 mg DNA and dissolve at 2 mg/mL in buffer I in a clean glass vial. Pass DNA 10 times through 21-gage needle into 5-mL syringe to ensure homogeneity of DNA (*see* **Note 8**).

2. Count (in duplicate at least) 10-µL aliquots of DNA solution in 3 mL Ecoscint.
3. Assume O^6-MeG content to be 6% of total cpm and set up ATase assay using up to 10-fold excess of ATase (*see* **Subheading 3.5.**). The plateau level will give an accurate measure of O^6-MeG content (calculate from specific activity of [^{3}H]MNU used). Plateau levels between 600 and 1200 cpm are appropriate for assaying most cells and tissues (*see* **Note 9**). DNA methylated using the above conditions gave 25 pmol O^6-MeG/mg DNA: 2 µg DNA was found to contain 580 cpm as ATase-repairable radioactivity (at 30% counting efficiency).
4. Carry out a kinetic experiment (i.e., incubate substrate and extract for increasing times) using the amount of extract giving 50% of the plateau level. This will determine the time for the reaction to go to completion.

3.2. Preparation of Extracts

Samples must be stored on ice throughout the procedure to conserve ATase activity.

1. Preparation and storage of cells/tissues/lymphocytes.
   Cells : harvest, wash with PBS, store cell pellet at –20°C.
   Tissue: snap-freeze (dry ice or liquid nitrogen) and store at –20°C or –70°C.
   Lymphocytes: collect whole blood into universal containing EDTA (final concentration 25 m*M*). Isolate lymphocytes by density centrifugation (**12**), wash with PBS, and store cell pellet at –20°C. The number of cells or weight of tissue required for the assay will depend on the level of ATase activity (e.g., *see* **Note 10**).
2. Transfer tissue or cell pellet to 1.5-mL Eppendorf tube in ice and add cold buffer I (500–1000 µL) containing 5 µg/mL leupeptin (*see* **Note 11**).
3. Sonicate sample (*see* **Note 3**) within a MSC class I cabinet to minimize exposure to aerosols. It may be necessary to mince with fine scissors or add glass beads to the sample to aid sonication (*see* **Note 12**).
4. Add PMSF to the sample immediately following sonication so that the final concentration is 0.5 m*M* (i.e., 1/100 of volume).
5. Centrifuge at 15–20,000*g* for 10 min at 4°C (*see* **Note 13**).
6. Transfer supernatant to a clean Eppendorf tube in ice. Extracts are now ready for use. For short-term storage (≤ 48 h), in ice, preferably in cold room/cabinet, is recommended. If for longer periods, extracts should be snap-frozen (dry-ice or liquid nitrogen) and stored at –20°C. Activity may be lost on freeze-thawing but we have not systematically investigated this.

3.3 Protein Estimation (13, see Note 14)

1. Add, in duplicate, 40 µL of each standard or unknown (*see* **Note 15**) to the bottom of a scintillation minivial. Blank tubes contain buffer I only.
2. Add 2 mL of CBG/ Bio-Rad reagent to each tube and gently mix (*see* **Note 5**).
3. Transfer 1 mL of blank sample to the cuvet and zero the machine at 595 nm. Repeat with duplicate blank sample to check reproducibility and stability of readings (*see* **Note 16**).
4. Transfer 1 mL of the unknown sample to the cuvet, record the reading, drain the cuvet thoroughly, and repeat with the next sample.

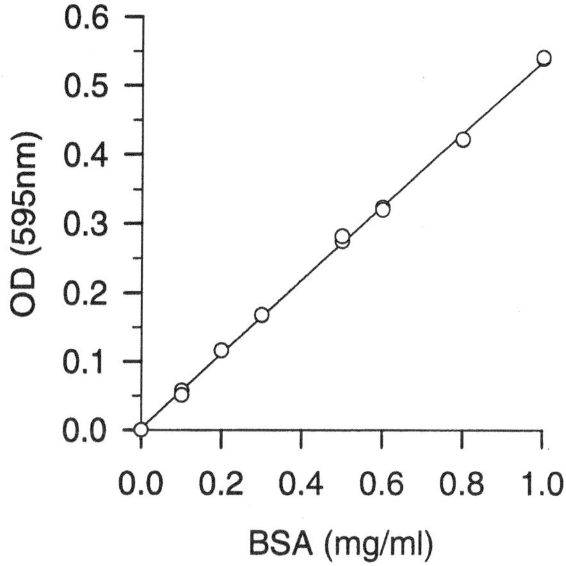

Fig. 1. Protein standard curve. Duplicate or triplicate sample readings are averaged and concentration is extrapolated from the standard curve. Standard deviation of protein estimations should be ≤5%.

5. Construct standard curve by plotting absorbance of standards vs protein concentration (mg/mL). Calculate protein concentration (mg/mL) of samplcs by reference to the standard curve (*see* **Fig. 1**, *see* **Note 17**).

3.4. DNA Estimation (14, see Note 6)

1. Switch on the TKO 100 at least 15 min before taking measurements so that the lamp is ready and the temperature in the chamber stabilizes.
2. Freshly prepare working solution of Hoechst 33258 by diluting to 1 µg/mL in 1X TNE, wrap in foil to protect from light, and allow to warm to room temperature before use.
3. Set the sensitivity of the detector monitor to about 50% by turning the scale knob approx 5 full clockwise turns from the fully counter position.
4. Add 2 mL of dye solution to the cuvet, if necessary wipe the sides of the cuvet with a low-lint tissue and place in the sample chamber.
5. Zero the instrument.
6. Deliver 2 µL of DNA standard into the 2 mL dye solution and mix by pipeting the solution into and out of a disposable pipet several times without introducing bubbles.
7. Close the sample chamber and adjust the scale control knob until the display reads "100." This one-point standardization sets the machine to read the DNA concentration directly in µg/mL. However, it is advisable to run a calibration

curve every few weeks to ensure that the standard has not degraded, the instrument is running properly, and that a consistent technique is being applied (*see* **Note 18**).

8. Remove the cuvet from the sample chamber. Empty and drain the cuvet thoroughly by blotting it upside down on a paper towel between readings.
9. Repeat **steps 4–8** at least once to verify that results are reproducible (*see* **Note 19**).
10. Read the DNA concentration (µg/mL) of your sample by repeating **steps 4–6**. Read the display immediately and record the value (*see* **Note 20**). Duplicate or triplicate reading are taken and averaged (*see* **Note 17**).

3.5. ATase Assay

1. Dilute substrate DNA to 200cpm/µL in buffer I (*see* **Subheading 3.1.3.**).
   At room temperature:
2. Dispense 100 µL (i.e., 20,000 cpm) of substrate into scintillation minivials taking care not to touch the sides (*see* **Note 21**).
3. Add appropriate volume of extract to the bottom of the tube, avoiding the substrate if possible. At least two "blank" tubes containing substrate only should be included to determine background counts (*see* **Note 21**).
4. Mix extract and substrate by gently shaking the tube taking care to avoid splashing the walls.
5. Add IBSA to the bottom of the tube to give a final volume of 300 µL (*see* **Note 22**).
6. Incubate at 37°C until the reaction is complete (*see* **Note 23**). Reaction time will depend on the particular substrate (*see* **Subheading 3.1.3.**) and tissue/cells used and should be established by experiment.
7. After incubation add 100 µL BSA, 100 µL 4 *M* PCA, 2 mL 1 *M* PCA. If reaction volume has been increased to 1.1 mL (*see* **Note 22**) add 400 µL 4 *M* PCA instead of 100 µL.
8. Heat at 75°C for 50 min to ensure complete hydrolysis of the DNA substrate (*see* **Note 21**).
9. Centrifuge at 2100*g* for 10 min.
10. Aspirate the supernatant (*see* **Note 24**) taking care not to disturb the pellet and add 4 mL PCA.
11. Repeat centrifugation and aspiration.
12. Add 300 µL 10 m*M* NaOH to each tube and shake to disperse the protein pellet, followed by 3 mL Ecoscint.
13. Cap tubes and shake thoroughly before scintillation counting.
14. Calculate the cpm for each tube by averaging two 5-min counts. Unusually high counts, particularly in the first few tubes, may indicate chemiluminescence and these tubes should be recounted to constant cpm.

3.6. Calculation of ATase Specific Activity

1. Plot cpm versus volume of extract for each sample and from the linear part of the curve calculate cpm/µL (*see* **Fig. 2**).
2. Multiply cpm/µL by conversion factor (*see* **Note 25**) to give fmoles/mL.

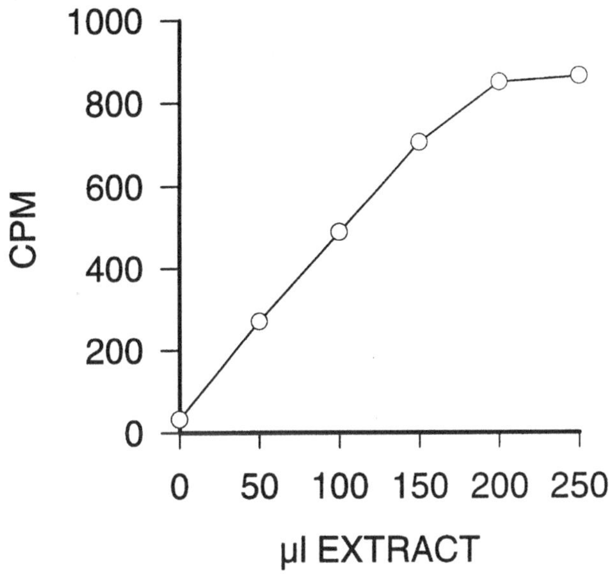

Fig. 2. Results of a typical ATase assay plotted as cpm versus volume of extract. Critical analysis of the data is important to ensure reliable results: (1) Background cpm should be within the range acceptable for the substrate (usually 20–50 cpm). (2) Minimum of three points must be linear (i.e., assay under protein-limiting conditions). (3) Cpm at maximum volume must be at least twice background cpm. (4) Standard deviation of assay ≤10%. (5) Plateau level of curve (in this case, 900 cpm) should correspond to the maximum number of cpm transferrable from substrate to protein (usually 6% of total cpm, *see* **Subheading 3.1.3.**). In some tissues, (e.g., rat GIT) where protease activity is high, plateau levels may be lower than expected presumably because of proteolytic degradation of the ATase. However, providing that specific activities are calculated based on the protein limiting (i.e., linear) part of the curve and all the above criteria are followed, reliable results may be obtained. (6) ATase specific activity of positive control sample (where used, *see* **Note 11**) is within acceptable range.

Assuming compliance with the foregoing steps, limit of sensitivity of this method ≈ ≥2fmoles ATase.

3. Divide fmol/mL by either protein (in mg/mL) or DNA (in µg/mL) concentration to give ATase specific activity in fmol/mg protein or fmol/µg DNA, respectively (*see* **Note 26**).

4. Notes

1. We routinely use Sigma calf thymus DNA (<3% protein).
2. Phenol is a powerful systemic poison, which is readily absorbed through the skin causing chemical burns on contact. Phenol should be neutralized with an equal volume of 4 *M* NaOH before disposal down the sink in high dilution.

3. We routinely use a Heatmaster Sonicator fitted with a microtip probe (standard tapered with 3.2-mm diameter) and sonicate for 10 s at 216 μm (peak-to-peak amplitude). Appropriate sonication conditions will depend upon the particular machine and probe used and should be established by experiment.
4. Bio-Rad dye reagent (Bio-Rad) supplied as a 5X stock may be used as an alternative to CBG. Standard curves using CBG are linear up to 1 mg/mL compared to those using Bio-Rad, which plateau around 0.6–0.8 mg/mL.
5. We recommend using an Eppendorf multipet plus (Eppendorf) or similar positive displacement machine to repeat dispense quickly and accurately.
6. Hoechst 33258 binds to the minor groove of DNA. When 365 nm light (long UV) excites this dye, fluorescence results and can be measured by mini-fluorometer (as described here), fluorescence spectrophotometer ***(15)***, or fluorescence microtiter plate-reader. We have also used a microtiter (96-well) based assay; 10 μL sample or standard (range 0–50 μg/mL DNA) and 100 μL Hoechst (1 μg/mL) per well.
7. If the interface is not clear, reextraction is necessary. Remove upper aqueous layer from each tube and combine in a Duran. Repeat **steps 3–5**.
8. It is important to ensure homogeneity of DNA by syringing because, quite often, the DNA forms small gelatinous "blobs," which are very difficult to see. If not dispersed these manifest themselves as highly variable counts on the plateau (substrate limiting conditions) of the assay.
9. 1It may be appropriate to use less substrate when trying to detect very low levels of ATase because less substrate will produce lower background counts. It is important that ATase-specific activity calculations are based on the linear part of the protein dependence curves.
10. Cells and tissues vary greatly in ATase activity and therefore the amount of extract required to detect activity will also vary. The following may be used as a starting guide if ATase activity is unknown.

    Cells: Bacterial—pelleted from 1.5-mL of stationary culture
    Mammalian—10^7 cells
    Lymphocytes—10 mL whole blood
    Tissue: 1–10 mg

    More (or less) cells or tissue may be used providing sonication is effective without compromising ATase activity.
    Inclusion of a positive control (i.e., ATase expressing cell line or tissue) in both extraction and assay procedures is strongly advised particularly when assaying samples with very low or unknown ATase activity.
    Occasionally, substrate-limiting conditions are attained at the lowest volume of extract used. In this case, repeat the assay using appropriate dilutions of extract in buffer I.
11. A minimum buffer volume of 500 μL is needed to avoid heating of the sample during sonication. If smaller extract volumes (300–500 μL) are required, it may be necessary to sonicate the sample for shorter times or on ice. Establish appropriate conditions by experiment.

12. Some tissues (e.g., gut and tumor) are more difficult to disrupt and therefore require the addition of 0.1–0.2 g glass beads (100 mesh, BDH) and/or mincing before sonication.
13. It may be possible to omit this step if very few cells are used, but this should be checked by experiment.
14. If a plate-reader with appropriate filter is available, this procedure may be adapted as a microtiter (96-well) plate assay. We have successfully used 40 µL of sample or standard (linear range 0–0.1 mg/mL protein) and 200 µL of dye reagent per well.
15. Samples (e.g., liver or tumor tissue) may need to be diluted in buffer I to achieve absorbance within linear range of standard curve (*see* **Fig. 1**).
16. Rezero the machine where necessary; frequency will depend on the particular machine.
17. Replicate protein and DNA estimations should be within 5%.
18. The DNA standard (100 µg/mL) is set to read 100, therefore a sample reading 35 has a concentration of 35 µg/mL. To generate a standard curve, follow **steps 1–9** (**Subheading 3.4.**), then measure (in duplicate) 4, 6, 8, and 10 µL of the DNA standard. Plot sample concentration vs the averaged reading. The data should yield a linear graph.
19. Once reproducible readings have been produced, do not touch the scale knob again until all measurements have been completed. Zero the instrument after every reading.
20. The observed fluorescence stabilizes after a few seconds and then begins to drop as the chamber warms.
21. High background counts (≥ twice normal background) usually indicate a problem with sample hydrolysis. Check hydrolysis incubation time and temperature, and PCA concentration. Ensure that one substrate is added to the bottom of the tube and is therefore effectively hydrolyzed.
22. The final volume is not critical providing the reaction goes to completion and the substrate is hydrolyzed in 1.0 *M* ± 5% PCA. For example, it may, occasionally be necessary to assay larger volumes (up to 500 µL) of cell or tissue extract in order to detect activity. In this case, the total assay volume may be increased to 1.1 mL with buffer I.
23. ATase is not a true enzyme, i.e., the reaction is stoichiometric and autoinactivating. Reaction should go to completion in all cases.
24. Dispose of waste in accordance with local rules for handling [^{3}H].
25. Calculation of conversion factor (CF).

    The conversion factor is calculated from the counting efficiency and substrate specific activity:

    a. dpm = cpm ÷ CE
       where CE = counting efficiency of scintillation counter (e.g., 30%)
    b. $1\ \text{Ci} = 2.22 \times 10^{12}$ dpm ∴ 1 pCi = 2.22 dpm
    c. Substrate specific activity (SA) in Ci/mmol (= pCi/fmol).
       Recalculate monthly by reference to decay chart

From a, b, and c.

fmol/μL	=	cpm/μL	×	$1/(SA \times 2.22 \times CE)$
fmol/mL	=	cpm/μL	×	$1000/(SA \times 2.22 \times CE)$
fmol/mL	=	cpm/μL	×	CF
∴ CF	=	$1000/(SA \times 2.22 \times CE)$		

26. Lymphocyte specific activities should be expressed in terms of fmol/μg DNA because in this way contaminating erythrocytes (which do not contain DNA) will not have a significant impact on the measurement. Although for most other purposes, either expression of activity is acceptable, there is a growing tendency to use fmol/μg DNA because from this value the number of ATase molecules per cell may be calculated.

References

1. Singer, B. (1979) *N*-nitroso alkylating agents: formation and persistence of alkyl derivatives in mammalian nucleic acids as contributing factors in carcinogenesis. *J. Natl. Cancer Inst.* **62,** 1329–1339.
2. Saffhill, R., Margison, G. P., and O'Connor, G. P. (1985) Mechanisms of carcinogenesis induced by alkylating agents. *Biochim. Biophys. Acta* **823,** 111–145.
3. Yarosh, D. B. (1985) The role of O^6-methylguanine-DNA methyltransferase in cell survival, mutagenesis and carcinogenesis. *Mutat. Res.* **145,** 1–16.
4. Margison, G. P. and O'Connor, P. J. (1990) Biological consequences of reactions with DNA: role of specific lesions, in *Handbook of Experimental Pharmacology,* vol. 94 (Grover, C. S. and Grover, P. L., eds.), Springer, Heidelberg, pp. 547–571.
5. Dolan, M. E., Moschel, R. C., and Pegg, A. E. (1990) Depletion of mammalian O^6-alkylguanine-DNA alkyltransferase activity by O^6-benzylguanine provides a means to evaluate the role of this protein in protection against carcinogenic and therapeutic alkylating agents. *Proc. Natl. Acad. Sci. USA* **87**, 5368–5372.
6. Lindahl, T., Demple, B., and Robbins, P. (1982) Suicide inactivation of the *E. coli* O^6-methylguanine- DNA methyltransferase. *Embo J.* **1,** 1359–1363.
7. Fan, C. Y., Potter, P. M., Rafferty, J., et al. (1990) Expression of a human O^6-alkylguanine-DNA alkyltransferase cDNA in human cells and transgenic mice. *Nucleic Acids Res.* **18,** 5723–5727.
8. Margison, G. P., Cooper, D. P., and Brennand, J. (1985) Cloning of the *E. coli* O^6-methylguanine and methylphosphotriester methyltransferase gene using a functional DNA repair assay. *Nucleic Acids Res.* **13,** 1939–1952.
9. Wilkinson, M. C., Potter, P. M., Cawkwell, L. C., et al. (1989) Purification of the *E.coli ogt* gene product to homogeneity and its rate of action on O^6-methylguanine, O^6-ethylguanine and O^4-methylguanine in dodecadeoxyribonucleotides. *Nucleic Acids Res.* **17,** 8475–8484.
10. Wu, R. S., Hurst-Calderone, S., and Kohn, K. W. (1987) Measurement of O^6-alkylguanine-DNA-alkyltransferase activity in human cells and tumour tissues by restriction endonuclease inhibition. *Cancer Res.* **47,** 6229–6235.

11. Brent, T. P. and Remack, J. S. (1988) Formation of covalent complexes between human *O*^{6}-alkylguanine-DNA alkyltransferase and BCNU-treated defined length synthetic oligonucleotides. *Nucleic Acids Res.* **16,** 6779–6788.
12 Coutinho, L. H., Gilleece, M. H., Dewynter, E. A., et al. (1993) Clonal and long-term culture using human bone-marrow, in *Haemopoesis, A Practical Approach* (Molyneux, G. M. and Testa, N. G., eds.), Oxford University Press, Oxford, UK, p. 288.
13. Bradford, M. M. (1976) A rapid and sensitive method for the quantitation of microgram quantities of protein utilising the principle of protein-dye binding. *Anal. Biochem.* **72,** 248–272.
14. Instructions—TKO 102 Standards Kit (Hoefer Scientific Instruments UK, Newcastle-under-Lyme, UK).
15. Cesarone, C. F., Bolognesi L., and Santi L. (1979) Improved microfluorimetric DNA determination in biological material using 33258 Hoechst. *Anal. Biol.* **100,** 188.

6

The *E. coli* Vsr Endonuclease

Assaying Activity In Vitro and In Vivo

Claire G. Cupples and Georgina Macintyre

1. Introduction

The primary function of the Vsr endonuclease of *Escherichia coli* is to initiate repair of T/G mismatches that arise from deamination of 5-methylcytosine to thymine (reviewed in *1*). By making a nick 5' of the T, Vsr provides a site for DNA polymerase I to start the removal of the mismatched base, along with several nucleotides on its 3' side. Resynthesis of the missing bases by the polymerase restores the correct C.G basepair and completes the process of very short patch (VSP) repair.

Strains of *E. coli* with defects in the *vsr* gene have a high frequency of C-to-T mutations *(2)*. These mutations occur at the second C of CCWGG sequences (W = A or T), the target site for Dcm, the sole DNA (cytosine-5) methyltransferase of *E. coli* K-12. T/G mismatches in this sequence context are, not surprisingly, the preferred substrate for Vsr *(3,4)*. Thus, in vitro assays for Vsr measure endonuclease activity in heteroduplexes containing a C(T/G)WGG sequence *(5)*, whereas, most currently used in vivo assays monitor C-to-T mutations at specific CCWGG sites in the *E. coli* genome *(6–8)*.

The in vitro assay for Vsr is relatively straight forward, simply requiring two oligonucleotides annealed to form the C(T/G)WGG site, the purified Vsr protein, and an appropriate incubation buffer *(3,5)*. There is no requirement for accessory proteins, and there are no apparent constraints on the DNA sequences that surround the Vsr target site. The oligonucleotide which contains the CTWGG sequence is labeled at the 5' or 3' end with ^{32}P. The single-stranded cleavage product is then detected by autoradiography following electrophoresis of the reaction mixture on a denaturing polyacrylamide gel.

From: *Methods in Molecular Biology, vol. 152: DNA Repair Protocols: Prokaryotic Systems*
Edited by: P. Vaughan © Humana Press Inc., Totowa, NJ

The Vsr protein is made in very small amounts by the cell *(9)*, and there is good evidence that it is unstable both in vitro *(3)* and in vivo *(9)*. Therefore, the easiest way to obtain large amounts of the protein is to clone the *vsr* gene into a multicopy plasmid, under the control of a strong promoter. One-step purification of Vsr is most efficiently achieved by fusing it to an easily purified moiety such as β-lactamase *(5)*, the maltose binding protein *(3)*, or an N-terminus hexahistidine tag. We describe the use of the His-tagged Vsr.

The original discovery of VSP repair came from studies of recombination between bacteriophage λ isolates in vivo *(10)*. Sequence analysis of closely linked genetic markers that recombined with unusually high frequencies revealed that one of the markers was always a CTAGG sequence. λ recombination assays were used for several years thereafter to dissect the genetic requirements of VSP repair *(11–13)*. However, in recent years, they have been replaced almost entirely by technically simpler assays that monitor the frequency of C-to-T mutations at specific CCAGG sites in the *E. coli* genome. In one assay, the mutation confers a change in the phenotype of the cell from kanamycin sensitivity to kanamycin resistance *(8)*. In two other assays *(6,7)*, mutation is accompanied by reversion from Lac^- to Lac^+. We will focus on one of the Lac reversion assays. It is highly specific for VSP repair, and may be used to screen populations of cells for those with defects in VSP repair *(6)*, or to determine the effect of deliberate alterations in the level of production, or activity, of Vsr and/or Dcm on mutation *(14,15)*. Its simplicity and low cost make it ideal for testing the activity of putative Vsr mutants before assaying them in vitro.

2. Materials

2.1. Reagents

All solutions are autoclaved unless otherwise noted. Water is double distilled.

1. Sonication buffer: 20 m*M* Tris-HCl, pH 8.0, 100 m*M* NaCl, adjust pH to 8.0.
2. Wash buffer: 20 m*M* Tris-HCl, pH 8.0, 100 m*M* NaCl, 10 m*M* imidazole, adjust pH to 8.0.
3. Vsr elution buffer: 20 m*M* Tris-HCl, pH 8.0, 100 m*M* NaCl, 200 m*M* imidazole, adjust pH to 8.0.
4. 2X formamide dye mix (10 mL): 1 mL 10X TBE, 9 mL deionized formamide, 50 mg bromophenol blue.
5. 10X TBE (1 L): 162 g Tris-base, 50 g boric acid, 7.5 g ethylenediaminetetraacetic acid (EDTA).
6. 15 % acrylamide gel stock (500 mL): 75 g acrylamide, 3.75 g N,N'-methylene-*bis*-acrylamide, 240 g urea, 50 mL 10X TBE, water. Add 0.54 mL 10% ammonium persulfate and 40 µL TEMED to 100 mL of gel solution just before use.

7. Gel elution buffer: 0.3 *M* NaCl, 10 m*M* Tris-HCl, pH 7.4, 1 m*M* EDTA, 1% phenol.
8. 10X kinase buffer: 0.5 *M* Tris-HCl, pH 7.6, 0.1 *M* $MgCl_2$, 50 m*M* dithrothreitol (DTT), 1 m*M* spermidine, 1 m*M* EDTA, pH 8.0.
9. TE: 10 m*M* Tris-HCl, pH 7.6, 0.1 m*M* EDTA, pH 8.0.
10. 20X SSC (1 L): 175.3 g NaCl, 88.2 g sodium citrate, 800 mL water, adjust pH to 7.0 with 10 N NaOH. Make to 1 L with water.
11. Reaction buffer: 25 m*M* HEPES-KOH, pH 7.9, 5 m*M* $MgCl_2$, 0.5 m*M* DTT.
12. Ampicillin: 100 mg/mL in water, filter sterilized. Concentration in media is 100 μg/mL.
13. LB medium (1 L): 10 g bacto tryptone, 10 g NaCl, 5 g bacto yeast extract. For plates, add 15 g agar before autoclaving.
14. Minimal A salts (10X) (1 L): 105 g K_2HPO_4, 45 g KH_2PO_4, 10 g $(NH_4)_2SO_4$, 5 g sodium citrate.$2H_2O$.
15. Minimal medium (1 L): Mix 100 mL 10X minimal A salts, 1 mL 1 *M* $MgSO_4.7H_2O$, 0.5 mL 1% thiamine, 10 mL 20% glucose or lactose. Make to 1 L with water. All components should be autoclaved before mixing. For plates, autoclave the water separately with 15 g agar, then add the other components.
16. Papillation plates. Make as for minimal glucose plates, adding 10 mL of 5% phenyl β-D-galactoside (Pgal) and 2 mL of 20 mg/mL Xgal (dissolved in N,N-dimethylformamide) to the 100 mL of salt solution.

2.2. Biologicals

1. Bacterial strains: CC112Δ *(15)—(ara* Δ*(gpt-lac)*5 *thi gyrA argE*(Am) *rpoB* Δ*(supD-dcm-fla)*, *zee*3129::*Tn*10, F' *lacZ-YA, proAB).* The strain also contains a glutamic acid-inserting, amber-suppressor gene, carried on a multicopy plasmid (Ap^R). BL21 (DE3) - [*hsdS, gal* (λ*c*I*ts*857 *ind*1 *S*am7 *nin*5 *lac*UV5-T7 gene*1*)].
2. Plasmids: pET-V–*vsr* gene cloned as an *Nde*I-*Bam*HI fragment in pET15b (Novagen, Inc., Milwaukee, WI) (Ap^R). pDV102 *(15)* - *Sph*I/*Bam*HI fragment containing *dcm* and *vsr*, with their own promoter, in pACYC184 (New England Biolabs, Beverly, MA) (Cm^R).

2.3. Supplies and Equipment

1. Whatman TLC plates with UV_{254} fluorescent, silica gel coating (cat. no. 4410–222) (Fisher Scientific).
2. Talon metal affinity resin (Clontech Laboratories).
3. Sephadex G50 nick spin columns (Amersham Pharmacia Biotech).
4. [γ-^{32}P]-ATP (3000 Ci/mmol) (Amersham Pharmacia Biotech).
5. T4 polynucleotide kinase (10 U/μL) (MBI Fermentas).
6. Glycogen (molecular biology grade) (Boehringer Mannheim / Roche).
7. Sequencing gel apparatus (e.g., Model S2 from Gibco-BRL).
8. Minivertical gel electrophoresis system (e.g., EC120 from E-C Apparatus Corp., supplied by Fisher Scientific).

3. Methods

3.1. *In Vitro* Assay of Vsr Endonuclease Activity

3.1.1. Purification of His-Tagged Vsr

The *his*-tagged Vsr is produced from *E. coli* strain BL21(DE3) transformed with plasmid pET-V. The plasmid contains the *vsr* gene, cloned in frame with six histidine codons, under the control of the T7 promoter. BL21(DE3) carries the gene for the T7 RNA polymerase expressed from the *lac* UV5 promoter. Induction of the polymerase with IPTG results in production of the *his*-tagged Vsr. The fusion protein is purified from cell extracts by binding it, via the hexahistidine tag, to a cobalt affinity resin. After several washing steps to remove unbound proteins, the fusion protein is removed from the resin by treatment with imidazole. Although the N-terminus hexahistidine tag may be removed from the Vsr by digestion with thrombin, this procedure is not necessary; the tagged protein is fully active.

1. Inoculate 20 mL of LB, containing ampicillin, with 0.2 mL of a fresh culture of BL21, transformed with pET-V. Grow at 37°C, with aeration, to an OD_{600} of 0.6. Induce production of His-tagged Vsr by adding 0.4 mL of 0.1 *M* IPTG. Grow for four more hours.
2. Harvest cells by centrifugation at 2000*g* for 5 min at 4°C. Resuspend the pellet in 2 mL of sonication buffer. Disrupt the cells with three freeze-thaw cycles; flash-freeze in dry-ice/ethanol; thaw in ice water.
3. Sonicate, with a microprobe, for four 15-s bursts separated by 30-s intermissions. Keep sample on ice during procedure.
4. Transfer 1.5 mL of sample to a microfuge tube. Centrifuge at 12,000*g* for 10 min at 4°C to pellet insoluble material. Transfer supernatant to a new tube. (Samples may be frozen at this point in dry ice/ethanol, and stored at –70°C.)
5. Prepare the Talon resin. Shake the resin until completely resuspended. Transfer 0.15 mL to a microfuge tube. (This volume contains enough resin to bind approx 0.4 mg of His-tagged protein.) Centrifuge at 700*g* for 2 min. Discard the supernatant. Add 0.75 mL of sonication buffer. Mix, repeat centrifugation, and discard supernatant.
6. Add sample to resin. Gently agitate at room temperature for 20 min. Centrifuge at 700*g* for 5 min. Remove supernatant.
7. Wash resin by adding 1.5 mL of wash buffer to the pellet. Gently agitate the suspension at room temperature for 10 min. Centrifuge at 700*g* for 5 min. Remove supernatant. Repeat wash step three times.
8. Elute His-tagged Vsr. Add 0.15 mL of elution buffer. Gently agitate suspension at room temperature for 10 min. Centrifuge at 700*g* for 5 min. Collect and save the supernatant. Repeat.
9. Check the two eluates for the presence of the His-tagged Vsr by electrophoresing 10 µL samples on an sodium dodecyl sulfate (SDS)-polyacrylamide gel, and staining the gel with Coomassie blue. The protein has an apparent molecular weight of approx 23 kDa (*see* **Note 1**).

*3.1.2. Preparation of DNA Substrates (see **Note 2**)*

3.1.2.1. Oligonucleotide Purification

Oligonucleotides can often be used as is, without further purification. However, purifying the oligonucleotides by gel electrophoresis before use eliminates shorter species that may obscure the Vsr reaction products.

1. Resuspend oligonucleotides in 200 µL water. Add 5 µL to 500 µL of water and measure the OD_{260}. Calculate the number of pmol/µL as follows: (OD_{260}/length of oligonucleotide) × 100 × dilution factor (100).
2. Mix 20 nmol of oligonucleotide, in 20 µL water, with 20 µL of 2X formamide dye mix. (If original oligonucleotide suspension is too dilute, concentrate in a vacuum dryer before use.) Heat to 90°C for 3 min. Chill on ice. Centrifuge briefly in microfuge.
3. Electrophorese on a 15% denaturing polyacrylamide gel, with 0.8-mm spacers, at 60 W until the bromophenol blue is half way down the gel. Place gel between two pieces of plastic wrap and place on a fluorescent thin-layer chromatography plate. Illuminate gel with short-wave ultraviolet light. DNA should show up as shadows on a fluorescent background. Cut full-length oligonucleotides out of gel with razorblade.
4. Put gel slices in microfuge tube with 1 mL of gel elution buffer. Place tube in 15-mL conical tube and incubate, with shaking, at 37°C overnight. Draw supernatant into a syringe, rinse tube with further 0.2 mL elution buffer, and filter total supernatant through a 0.2-micron Millipore filter.
5. Add 2 µL of 20 µg/µL glycogen and 10 µL of 1*M* $MgCl_2$. Divide solution between two microfuge tubes and add 1/10 volume 3 *M* NaAc, pH 5.2, and 2 vol of 95 % EtOH to each tube. Mix and incubate at –20°C overnight.
6. Centrifuge at 12,000*g* for 30 min at 4°C. Vacuum dry sample, combine the contents of each tube in 25 µL of water. Calculate the concentration as described above, and adjust to 10 pmol/µL. Store purified oligonucleotides at –20°C.

3.1.2.2. 5' End-Labeling and Annealing of Oligonucleotides

End-label the two oligonucleotides that form the top strands of the hetero- and homoduplexes (*see* **Note 2**). A separate oligonucleotide, the size of the expected Vsr cleavage product, should also be end-labeled. All procedures involving ^{32}P should be done following the appropriate institutional and/or governmental guidelines regulating the use of radioisotopes.

1. Mix the following in a microfuge tube: 2 µL (20 pmol) of oligonucleotide, 2 µL 10X kinase buffer, 2 µL [γ-^{32}P]-ATP, 1 µL T4 polynucleotide kinase, water to 20 µL. Incubate at 37°C for 45 min. Stop the reaction by incubating at 70°C for 10 min.
2. To remove unincorporated label, use a Sephadex G-50 nick spin column. First, prepare the column. Resuspend the Sephadex G-50 by inverting the column several times. Remove caps from the column—top first, then bottom. Allow column to drain. Add 1 mL 1X TE and allow column to drain. Place column in a plastic

15-mL centrifuge tube and add 2 mL 1X TE. Allow column to drain. Centrifuge at 500*g* in a swinging bucket rotor. Discard the eluate.

3. Next apply the radioactive sample, in 100 µL 1X TE, to the column. Place the column in a clean 15-mL centrifuge tube containing an uncapped microfuge tube; the tip of the column should sit in the microfuge tube. Elute the sample by centrifugation at 500*g* for 4 min. Remove the microfuge tube, cap it, and store at –20°C.
4. To anneal the oligonucleotides, mix the following: 10 pmol radioactively labeled oligonucleotide (half total eluate), 5 µL (50 pmol) unlabeled oligonucleotide, 25 µL 20X SSC, water to 500 µL. Incubate at 80°C for 5 min, then slowly cool to room temperature. (Estimated concentration of annealed product: 40 fmol/µL.) Store at –20°C.

3.1.3. Assaying Vsr Activity

Assay Vsr activity against four DNA substrates: two heteroduplexes, one with a T/G mismatch and the other with a C/A mismatch, and two homoduplexes, one with a T.A basepair and the other with a C.G basepair. Vsr should only cleave the T/G heteroduplex. We have detected activity from as little as 6 ng of Vsr.

1. To ensure consistency among samples, prepare a master mix for each pair of annealed oligonucleotides, sufficient for the planned number of reactions: mix oligonucleotides (diluted to 10 fmol/µL), 10X reaction buffer and water in a 1:1:7 ratio.
2. For each reaction, combine 18 µL of master mix and 2 µL (100 ng) of purified Vsr. Control reactions, with no Vsr, should be included for each oligonucleotide pair. Incubate tubes at 30°C for 15 min.
3. Add 20 µL phenol:chloroform (1:1), vortex, and centrifuge for 2 min. Transfer 10 µL of supernatant to new tubes containing 10 µL 2X loading dye. At the same time, prepare a tube with an equivalent amount of the end-labeled size marker, mixed with an equal volume of 2X loading dye. Heat samples to 90°C for 3 min, then chill immediately in ice water. Load 10 µL onto pre-run, 15% denaturing polyacrylamide gel (**Note 3**). Electrophorese at 60 W for about 3 h, or until bromophenol blue has migrated 3/4 of the way down the gel. Dry gel, and place in film cassette with X-ray film and intensifying screens. Expose at –80°C.

3.2. In Vivo Assay of Vsr-Mediated Mismatch Repair

3.2.1. Overview

In vivo assays for Vsr measure the ability of the enzyme to prevent C-to-T mutations caused by deamination of 5-methylcytosine to thymine. Decreased repair activity is accompanied by increased mutation. In the Lac reversion assay described here, a CCAGG to CTAGG mutation in the *lacZ* gene of *E. coli* strain CC112 converts a glutamine codon (CAG) at position 461 to a nonsense codon (TAG). In the presence of an amber-suppressor gene carried on a multicopy plasmid, this codon is translated as glutamic acid, the only amino acid at that site that will give a functional β-galactosidase *(16)*. The result is a change in the phenotype of the cell from Lac^- to Lac^+. Although a glutamic acid codon can arise directly from a C-to-G transversion, this is a very rare mutation in *E. coli*. Therefore, this assay is highly specific for VSP repair (**Note 4**).

For assaying Vsr mutants, we use a *vsr* version of CC112. CC112Δ has a chromosomal deletion that removes both the *vsr* gene and the *dcm* gene. The two genes, wild type or mutant, are restored on a plasmid. The inclusion of *dcm* is essential because methylase activity is necessary to produce the lesion that Vsr repairs (i.e., the T/G mismatch that results from deamination of 5-methylcytosine to thymine). In addition, the 5' end of *vsr* overlaps the 3' end of *dcm* on the *E. coli* chromosome, so expression of *vsr* depends on a promoter 5' of *dcm (17)*. (*See* **Note 5** for alternative ways of expressing *vsr*.) The *dcm-vsr* plasmid must contain a p15a origin of replication because the suppressor plasmid has a colE1 origin; pACYC184 derivatives such as pDV102 (**Fig. 1**) are a good choice.

3.2.2. The Papillation Assay

Increased or decreased Lac reversion is scored by plating CC112Δ transformants on papillation medium and noting the relative numbers of blue papillae in individual colonies. The plates contain glucose, Xgal, and either lactose or Pgal *(18)*. When cells exhaust the supply of glucose, the Lac^- parental cells cease to grow. Lac^+ revertants continue to grow using the Pgal as a carbon source. The papillae that they form, stained blue by hydrolysis of Xgal, are easily seen against the white background colony. For example, **Fig. 1** shows that colonies arising from cells producing enzymatically inactive Vsr's (pKD1 and pKD2) have many more papillae than a colony derived from a cell that is producing the wild-type enzyme (pDV102). The number of papillae is comparable to that seen in a cell that is producing Dcm, but no *Vsr* (pDV101).

1. Inoculate 1 mL of LB medium with a single colony of CC112Δ transformed with *vsr*-containing plasmid.
2. Incubate at 37°C for several hours, without aeration, until culture starts to become turbid (**Note 6**).
3. Add 100 µL of this culture to 2 mL of minimal glucose medium, containing ampicillin and the appropriate antibiotic for the *vsr*-containing plasmid.
4. Incubate at 37 °C, with aeration, until culture is saturated (at least overnight).
5. Spot 5 to 10 µL of undiluted culture onto fairly dry papillation plates, forming an artificial colony. Let the spot dry completely.
6. Incubate plates at 37°C for several days, until "colonies" are well grown, and dark blue papillae are visible.

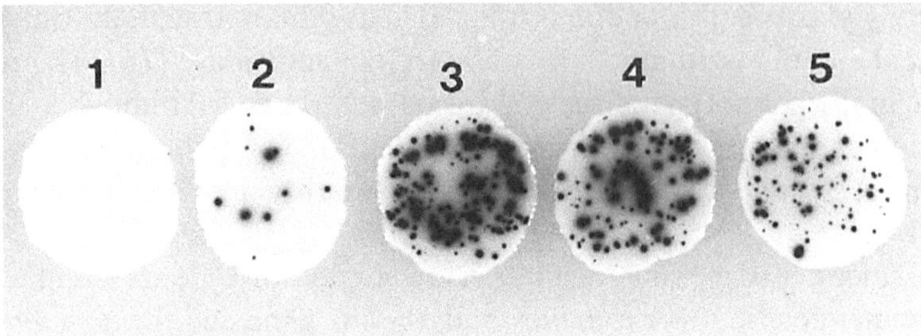

Fig. 1. The papillation assay for Vsr activity. CC112Δ transformed with 1: pACYC184 (control); 2: pDV102 (*dcm*, wild-type *vsr*); 3: pKD1 (*dcm*, mutant *vsr* #1); 4: pKD2 (*dcm*, mutant *vsr* #2); 5: pDV101 (*dcm* alone). (Doiron, K. M. J. and Cupples, C. G., unpublished data.)

3.2.3. The Assay on Minimal Lactose Plates

For a more quantitative assessment of repair activity, plate cells on minimal lactose (ML) medium instead of papillation medium.

1. Follow **steps 1–4** of the papillation assay.
2. Spread 100 µL of a 10^{-6} dilution of the culture onto LB plates (to assess viability), and 100 µL of undiluted culture onto minimal lactose plates. If screening multiple cultures simultaneously, spot 10 µL quantities of undiluted culture on ML plates, and 10 µL quantities of diluted culture on LB plates.
3. Incubate the LB plates overnight, and the minimal lactose plates for at least 36 h, both at 37°C.
4. Count the colonies, and determine the number of Lac^+ mutants/10^8 viable cells.

4. Notes

1. There is one published report ***(3)*** that Vsr is unstable in vitro, readily losing 12–14 amino acids from the N terminus. We have, on occasion, observed both the long and the short species in vivo, the long version being the predominant one. Although the Δ14 species has little, if any activity in vivo ***(17)***, it appears to have full activity in vitro ***(3)***. However, the relative activity of the two species has not, to our knowledge, been explicitly tested.

   We have found that the *his*-tagged Vsr is very stable. The protein remains intact for at least two weeks at 4°C.
2. We use oligonucleotides of 31 bases that form blunt-ended fragments when annealed. We make four oligonucleotides to allow for the construction of two homoduplexes: C(C/G)AGG and C(T/A)AGG), and two heteroduplexes: (C(T/G)AGG and C(C/A)AGG). Only the C(T/G)AGG heteroduplex should cut with Vsr; the other three provide negative controls. Examples of successful oligonucleotide are shown here (X = C or T; Y = G or A).

Top strand: 5' - GCGGCGGGCCATTA<u>CXAGG</u>CCGAAGCAGCGT - 3'
Bottom strand: 5' - ACGCTGCTTCGG<u>CCTYG</u>TAATGGCCCGCCGC - 3'

In this example, Vsr cleavage of the T/G-containing heteroduplex, labeled at the 5' end of the top strand, yields a 15 base product. Therefore, we electrophorese a 5'-end-labeled 15 mer in parallel with the reaction mixtures, to serve as a size marker.

3. For optimal resolution of the reaction products, electrophorese the samples on a 0.2-mm-thick, 40-cm-long sequencing gel. However, for quick analysis, one can obtain adequate resolution of 30 mers and 15 mers on a minivertical gel electrophoresis system (8 cm in length).
4. CC112 is one of a series of *E. coli* strains designed to monitor the frequency of occurrence of specific mutations. Like CC112, CC101–106 ***(19)*** each have a missense or nonsense codon in place of the wild-type glutamic acid codon at 461 in *lacZ,* which makes them Lac^-. Reversion to Lac^+ in each strain requires a unique base substitution that restores the glutamic acid codon (GAG). CC107–111 require frameshift mutations in *lacZ* to revert from Lac^- to Lac^+ ***(20)***. In all 11 strains, the *lacZ* alleles are carried on F episomes along with *lacY, lacA,* and *proAB*; the chromosomal *lac* operon is deleted. The episomes can be transferred to any *lacZ, proAB* strain of *E. coli* by conjugation ***(18)***, selecting for proline prototrophy. Prior deletion of chromosomal *lacZ, proAB* in the recipient strain is done by conjugation with CSH63 (*val*^{R} *thi* Δ(*gpt-lac*)5, HfrH), or by P1 transduction using a CSH63 lysate and selecting for tetracycline or valine (50 µg/mL L-valine) resistance.
5. Expression of *vsr* independent of *dcm* can be achieved by cloning the gene under the control of a synthetic promoter such as *trc* ***(14)***, and introducing the plasmid into a Dcm^+, Vsr^- strain, such as CC112V. This strain has a kanamycin resistance cassette in *vsr,* which eliminates Vsr without affecting production of the methylase ***(21)***. However, overexpression of Vsr causes mutation at non-CCWGG sequences. Vsr-stimulated mutagenesis is almost certainly caused by interference with mismatch repair; in fact, the types of mutation that occur (transitions and frameshifts) and the frequency with which they occur are comparable to that seen in cells defective in one or other of the mismatch repair genes ***(15)***.

High expression, *vsr*-containing plasmids can actually be used as a tool for generating random mutations in any gene, native or foreign, in *E. coli*. In effect, transformation with one of the plasmids makes any strain of *E. coli* into a mutator. Unlike the chemical mutagens or ultraviolet radiation that are usually used to produce mutations in *E. coli*, *vsr*-containing plasmids are safe to work with, and do not generate hazardous waste. If the target gene is carried on a plasmid, the *vsr*-containing plasmid must have a compatible origin of replication. pKK-V ***(14)*** has a colE1 origin of replication. pDV108 has a p15a origin; it contains the *vsr* gene with the trc promoter subcloned as a *Hin*dIII-*Bam*HI fragment from pKK-V into pACYC184.

6. The CC112Δ strain has an amber mutation in the *argE* gene. Growth of the cells in minimal medium, containing ampicillin, selects for the maintenance of the plasmid carrying the amber suppressor tRNA gene. However, cells grow best when they are cultured in LB for a short while before being introduced into the minimal medium.

   If experiments require addition of two plasmids (e.g., separate clones of *dcm* and *vsr*), use the CC462 strain instead of CC112 *(7)*. In this strain, the CCAGG to CTAGG mutation converts a proline codon (CCA) at position 462 in *lacZ* into a leucine codon (CTA), restoring β-galactosidase function. Although other base substitution mutations that result in nonproline codons will also give a functional protein, in practice the majority of mutations are the desired C-to-T transition. If using CC462 on ML medium, remove the plates after no more than 36 h to avoid high background. For papillation assays, use MacConkey lactose plates ***(18)*** because the light blue color of this strain on Xgal plates makes it difficult to see the dark blue papillae.

Acknowledgments

We thank Kathy Doiron for **Fig. 1.** Research in this laboratory has been supported by grants to CGC from the Natural Sciences and Engineering Research Council of Canada, the Medical Research Council of Canada, and the National Cancer Institute of Canada (with funds from the Canadian Cancer Society).

References

1. Lieb, M. and Bhagwat, A. S. (1996) Very short patch repair: reducing the cost of cytosine methylation. *Mol. Microbiol.* **20,** 467–473.
2. Lieb, M. (1991) Spontaneous mutation at 5-methylcytosine hotspot is prevented by very short patch (VSP) mismatch repair. *Genetics* **128,** 23–27.
3. Gläsner, W., Merkl, R., Schellenberger, V., and Fritz, H.-J. (1995) Substrate preferences of Vsr DNA mismatch endonuclease and their consequences for the evolution of the *Escherichia coli* K-12 genome. *J. Mol. Biol.* **245,** 1–7.
4. Lieb, M. and Rehmat, S. (1995) Very short patch repair of T:G mismatches in vivo: Importance of context and accessory proteins. *J. Bacteriol.* **177,** 660–666.
5. Hennecke, F., Kolmar, H., Bründl, K., and Fritz, H.-J. (1991) The *vsr* gene product of *E. coli* K-12 is a strand- and sequence-specific DNA mismatch endonuclease. *Nature* **353,** 776–778.
6. Ruiz, S. M., Létourneau, S., and Cupples, C. G. (1993) Isolation and characterization of an *Escherichia coli* strain with a high frequency of C-to-T mutations at 5-methylcytosines. *J. Bacteriol.* **175,** 4985–4989.
7. Petropoulos, L., Vidmar, J. J., Passi, E., and Cupples, C. G. (1994) A simple assay for monitoring the mutagenic effects of 5-methylcytosine deamination in *Escherichia coli. Mutat. Res.* **304,** 181–185.

8. Wyszynski, M., Gabbara, S., and Bhagwat, A. S. (1994) Cytosine deaminations catalysed by DNA cytosine methyltransferases are unlikely to be the cause of mutational hot-spots at the sites of cytosine methylation in *E. coli. Proc. Natl. Acad. Sci. USA* **91,** 1574–1578.
9. Macintyre, G., Pitsikas, P., and Cupples, C. G. (1999) Growth phase-dependent regulation of Vsr endonuclease may contribute to 5-methylcytosine mutational hotspots in *Escherichia coli. J. Bacteriol.* **181,** 4435,4436.
10. Lieb, M. (1983) Specific mismatch correction in bacteriophage lambda crosses by very short patch repair. *Genetics* **191,** 118–125.
11. Lieb, M. (1987) Bacterial genes *mutL, mutS* and *dcm* participate in repair of mismatches at 5-methylcytosine sites. *J. Bacteriol.* **189,** 5241–5246.
12. Jones, M., Wagner, R., and Radman, M. (1987) Mismatch repair of deaminated 5-methylcytosine. *J. Mol. Biol.* **194,** 155–159.
13. Dzidic, S. and Radman, M. (1989) Genetic requirements for hyper-recombination by very short patch repair: involvement of *Escherichia coli* DNA polymerase I. *Mol. Gen. Genet.* **217,** 254–256.
14. Doiron, K. M. J., Viau, S., Koutroumanis, M., and Cupples, C. G. (1996) Overexpression of *vsr* in *Escherichia coli* is mutagenic. *J. Bacteriol.* **178,** 4294–4296.
15. Macintyre, G., Doiron, K. M. J., and Cupples, C. G. (1997) The Vsr endonuclease of *Escherichia coli*: an efficient DNA repair enzyme and a potent mutagen. *J. Bacteriol.* **179,** 6048–6052.
16. Cupples, C. G. and Miller, J. H. (1988) Effects of amino acid substitutions at the active site in *Escherichia coli* β-galactosidase. *Genetics* **120,** 637–644.
17. Dar, M. E. and Bhagwat, A. S. (1993) Mechanism of expression of DNA repair gene *vsr*, an *Escherichia coli* gene that overlaps the DNA cytosine methylase gene, *dcm. Molec. Microbiol.* **9,** 823–833.
18. Miller, J. H. (1992) *A Short Course in Bacterial Genetics.* Cold Spring Harbor Laboratory Press, Cold Spring Harbor, New York.
19. Cupples, C. G. and Miller, J. H. (1989) A set of *lacZ* mutations in *Escherichia coli* that allow rapid detection of each of the six base substitution mutations. *Proc. Natl. Acad. Sci. USA* **86,** 5345–5349.
20. Cupples, C. G., Cabrera, C., Cruz, C., and Miller, J. H. (1990) A set of *lacZ* mutations in *Escherichia coli* that allow rapid detection of specific frameshift mutations. *Genetics* **125,** 275–280.
21. Doiron, K. M. J., Lavigne-Nicolas, J., and Cupples, C.G. (1999) Effect of interaction between 5-azacytidine and DNA (cytosine-5) methyltransferase on C-to-G and C-to-T mutations in *Escherichia coli. Mutat. Res.* **429,** 37–44.

7

Coupling of DNA Helicase Function to DNA Strand Exchange Activity

Frank G. Harmon and Stephen C. Kowalczykowski

1. Introduction

DNA repair can occur by a variety of mechanistically distinct pathways [for review, *see* (*1*)]. Recombinational DNA repair is one such pathway, and it requires the coordinated action of many different enzymes. In the best studied organism, *Escherichia coli*, more than 20 different proteins are involved [for review, *see* (*2*)]. The recombinational repair of a double-stranded DNA (dsDNA) break requires four general steps: (1) processing; (2) homologous pairing; (3) DNA heteroduplex extension; and (4) resolution. Here we describe assays to study aspects of the first two steps.

One of the most important proteins for DNA repair by homologous recombination is the RecA protein [for review, *see* (*3–5*)]. RecA-like proteins are highly conserved and are present in all organisms thus far examined *(6)*. Loss of RecA protein function renders *E. coli* cells highly sensitive to DNA damaging agents *(7)*, and reduces recombination frequency to 0.1% of wild-type levels *(8)*. In vitro, RecA protein promotes both the pairing and exchange of DNA strands between two homologous DNA molecules, provided that one of them contains a region of single-stranded DNA (ssDNA). RecA protein assembles on this ssDNA to form a nucleoprotein filament, referred to as the presynaptic complex. It is this nucleoprotein filament that is active for homologous pairing and DNA strand exchange *(9)*.

In vivo, most DNA breaks do not directly produce a ssDNA substrate for RecA protein, thus necessitating processing of the lesion by other enzymes *(1)*. In most cases, damage-specific DNA helicases process the DNA break to produce a region of ssDNA sufficient for binding by RecA protein *(2)*. In *E. coli*, dsDNA breaks are acted upon by RecBCD enzyme, a DNA helicase/nuclease

From: *Methods in Molecular Biology, vol. 152: DNA Repair Protocols: Prokaryotic Systems*
Edited by: P. Vaughan © Humana Press Inc., Totowa, NJ

(2,10–13). If, instead, the lesion produces a ssDNA break or a ssDNA gap, processing is provided by the RecQ helicase *(14–16)*. Thus, the initial steps of recombinational DNA repair require that the unwinding activity of a DNA helicase is coupled to the homologous pairing and DNA strand exchange activity of the pairing protein, RecA protein.

The coupling of RecA protein and a DNA helicase to promote homologous pairing between homologous dsDNA has been achieved in vitro for both RecBCD enzyme *(17,18)* and RecQ helicase *(16)*. The general approach is to use DNA components that are substrates for the DNA helicase, but not for RecA protein, unless the DNA is first processed by the helicase. The assays use ^{32}P-labeled, linear dsDNA as the ssDNA-donor and a homologous supercoiled DNA (scDNA) as the recipient for homologous pairing (**Fig. 1A**). Because RecA protein alone is unable to pair the two fully dsDNA substrates *(19)*, this assay requires that the linear dsDNA substrate be separated into its constituent ssDNA strands so that RecA protein can form the presynaptic complex that is necessary for homologous pairing and DNA strand exchange. RecA protein-promoted homologous pairing between the linear ssDNA and its scDNA homolog produces a paired DNA species known as a joint molecule. The joint molecule consists of the linear ssDNA molecule (produced by dsDNA unwinding) basepaired to a complementary region within the recipient scDNA molecule (**Fig. 1A**). In the joint molecule, homologous pairing between the ssDNA and scDNA molecules results in the production and displacement of an unpaired DNA strand; therefore, this type of joint molecule is referred to as a displacement-loop or D-loop.

We describe two reaction protocols for the formation of joint molecules: one designed to test an uncharacterized RecA-like protein for homologous pairing activity (partial reaction) and the second designed to couple this activity to the unwinding activity of a DNA helicase (coupled reaction). In both cases, the two DNA substrates are pUC19 scDNA and ^{32}P-labeled, linear pUC19 dsDNA. The latter is produced by digestion of pUC19 scDNA with a restriction enzyme, followed by 5'-end labeling with (γ-[^{32}P]) ATP using T4 polynucleotide kinase. The two protocols differ only in the technique used to produce the linear ssDNA substrate for RecA protein. In the partial reaction, RecA protein is provided with linear ssDNA, which is generated by heat-denaturation to obviate the need for a DNA helicase. The coupled reaction, having linear dsDNA and scDNA as the DNA substrates, requires a DNA helicase to unwind the linear dsDNA to furnish RecA protein with a ssDNA substrate. Included in the protocol are three separate buffer conditions to optimize the reaction for uncharacterized proteins. The products of these reactions are analyzed by agarose gel electrophoresis, where D-loops appear as ^{32}P-labeled DNA species with a retarded mobility relative to the ^{32}P-labeled, linear ss- or dsDNA sub-

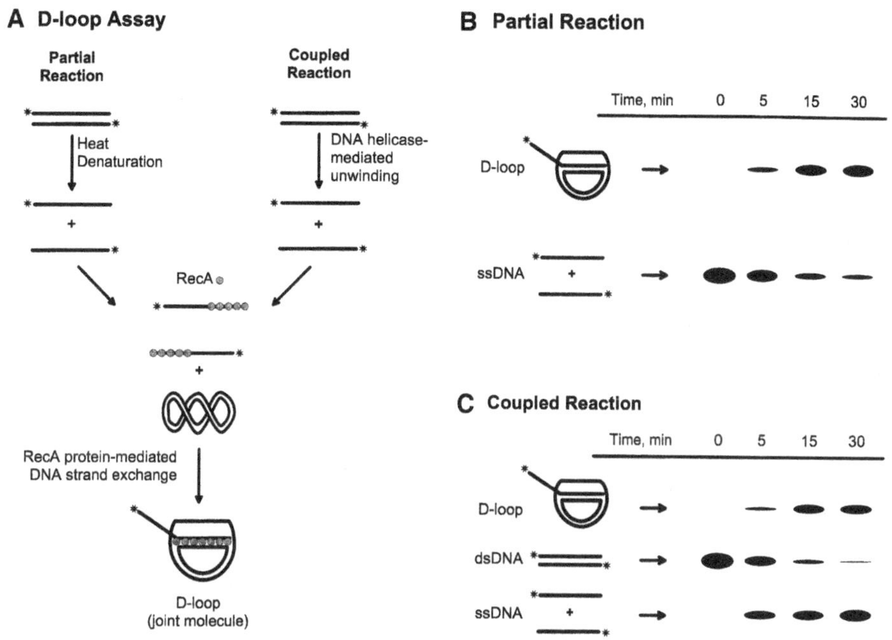

Fig. 1. Schematic diagram of the homologous pairing assays. See text for details. (**A**) D-loop assay for both the partial (i.e., RecA protein alone) and coupled (i.e., both RecA protein and a DNA helicase) assays. (**B**) Typical time-course of D-loop formation mediated by RecA protein in the partial reaction. (**C**) Typical time-course for a coupled reaction having both RecA protein and a DNA helicase.

strates (representations of typical autoradiograms are shown in **Fig. 1B and C**). In addition, we provide an agarose gel electrophoresis-based helicase assay, first described by Matson et al. *(20)*, to determine the specific activity of an uncharacterized DNA helicase (**Fig. 2**). The specific activity calculated from this assay can then be used to determine the amount of the DNA helicase to use in the coupled reaction. As with the homologous pairing assays, the helicase assay protocol includes three separate buffer conditions to optimize unwinding for an uncharacterized DNA helicase.

2. Materials

2.1. Stock Solutions

All reagents were from Sigma-Aldrich (Boston, MA), except where noted.

1. 1 *M* Tris acetate, pH 7.5.
2. 0.1 *M* magnesium acetate, pH 7.5.
3. 0.1 *M* dithiotheritol (DTT) in water. Store aliquots at –20°C.
4. 0.5 *M* ethylenediaminetetraacetic acid (EDTA), pH 8.0.

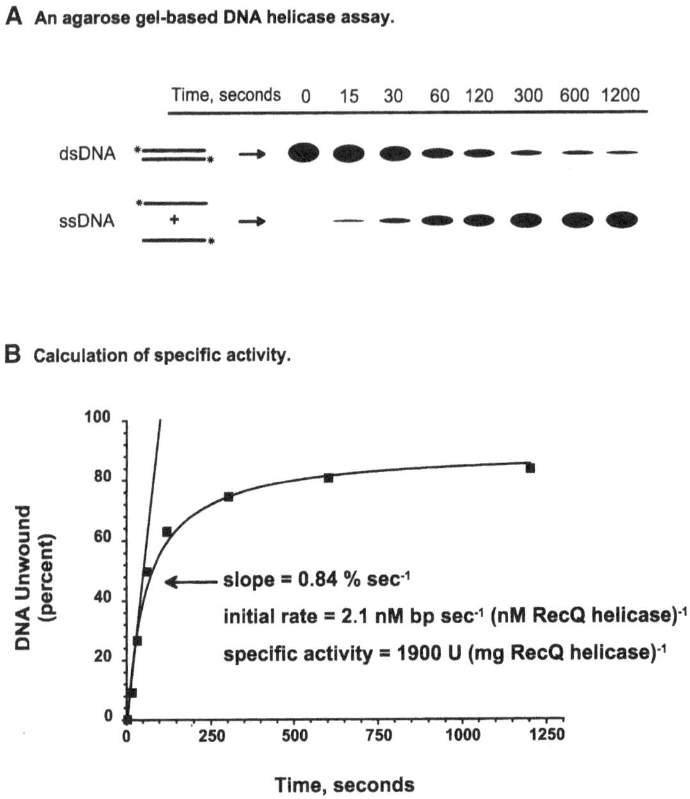

Fig. 2. DNA helicase activity assay. **(A)** Typical time-course for the unwinding of 10 µM (nucleotides) linear, pUC19 dsDNA by 20 n*M* RecQ helicase using agarose gel electrophoresis to detect strand separation. **(B)** Calculation of the specific activity of RecQ helicase from the data in A. *See* **Subheading 3.2.** for details.

5. 10% sodium dodecyl sulfate (SDS).
6. 0.1 *M* ATP in water, adjusted to pH 7.0 with NaOH. Store aliquots at –20°C.
7. 0.1 *M* phoshpoenol pyruvate (PEP) adjusted to pH 7.0 with NaOH. Store aliquots at –20°C.
8. Pyruvate kinase: Pipet 100 U of pyruvate kinase solution (homogeneous $(NH_4)_2SO_4$ suspension from rabbit muscle) into a 1.5-mL microcentrifuge tube, pellet protein by centrifugation in microcentrifuge (Micromax, IEC, Needham Heights, MA) at 20,000*g* for 2 min at 4°C, remove the supernatant, and dissolve the protein pellet in 50 µL of 1X pairing buffer. Final concentration is 2000 U/mL. Store at 4°C for up to 2 wk.
9. 40 % (w/v) polyethylene glycol (PEG): Dissolve 40 g of dry PEG (8000 molecular weight) in 40 mL water and mix well. Bring volume to 100 mL with water. Protect from light and store at room temperature.

10. 5X stop buffer: 250 m*M* EDTA, and 5% SDS. Store at room temperature for up to 1 wk.
11. 5X sample loading buffer: 15 % (w/v) Ficoll (Amersham-Pharmacia Biotech, Piscataway, NJ), in water with 0.25 % (w/v) each of xylene cyanol and bromophenol blue. Store at room temperature.
12. Stop solution: Mix equal volumes of 5X stop buffer and 5X sample loading buffer, and add proteinase K (Roche-Boehringer Mannheim) to 4.5 mg/mL. Prepare immediately before use and discard unused portion.
13. 1X TAE buffer for agarose electrophoresis: 40 m*M* Tris acetate, pH 8.0, 1 m*M* EDTA. Prepared as 50X concentrated solution: 2 *M* Tris base, 1 *M* acetic acid and 50 m*M* EDTA.
14. Deionized water using NANOpure reagent grade water system (Barnstead-Thermolyne, Dubuque, IA).
15. Agarose: type II, low EEO.
16. DE81 anion exchange paper (Whatman).

2.2. Preparation of 5'-End-Labeled Linear DNA Substrate

1. DNA: 2–10 µg of pUC19 scDNA. This DNA can be purchased from a commercial source or purified as described ***(21)***.
2. Restriction endonuclease (*Hind*III) and T4 polynucleotide kinase (New England Biolabs).
3. Shrimp alkaline phosphatase (United States Biochemical).
4. 10X restriction buffer: 100 m*M* Tris-HCl, pH 7.5, 100 m*M* $MgCl_2$, 1000 m*M* NaCl, 10 m*M* DTT. Store aliquots at –20°C.
5. MicroSpin G-25 columns (Amersham Pharmacia Biotech).
6. γ^{32}P ATP, 1000 µCi/mL.

2.3. DNA Helicase Assay

1. DNA helicase of interest. Purification of *E. coli* RecQ helicase and *E. coli* RecBCD enzyme have been described previously ***(11,16)***.
2. SSB protein (*see* **Note 1**). *E. coli* protein is commercially available: Promega and United States Biochemical. Purification of *E. coli* SSB protein has been described previously ***(22)***.
3. DNA: 2 µg 5'-end-labeled linear pUC19 dsDNA at a stock concentration of 250 µM (nucleotides) (*see* **Note 2**).
4. 10X helicase buffer: 250 m*M* Tris acetate, pH 7.5, 10 m*M* ATP, 10 m*M* PEP, and 10 m*M* DTT. Store aliquots at –20°C.

2.4. RecA Protein-Mediated D-Loop Formation—Partial Reaction

1. RecA-like protein of interest. Purification of *E. coli* RecA protein has been described previously ***(23,24)***. RecA protein is also commercially available: Promega, United States Biochemical, New England Biolabs.
2. SSB protein (*see* **Notes 1** and **3**). *E. coli* protein is commercially available: Promega and United States Biochemical. Purification of *E. coli* SSB protein has been described previously ***(22)***.

3. DNA: 2 μg 5'-end-labeled linear pUC19 ssDNA and 4 μg pUC19 scDNA, at a stock concentration of 250 and 500 μM (nucleotides), respectively (*see* **Note 2**).
4. 10X pairing buffer: 250 m*M* Tris acetate, pH 7.5, 10 m*M* ATP, 10 m*M* PEP, and 10 m*M* DTT. Store aliquots at –20°C.

2.5. Coupling DNA Strand Exchange and a DNA Helicase to Form D-Loops—Coupled Reaction

1. RecA-like protein of interest. Purification of *E. coli* RecA protein has been described previously *(23,24)*. RecA protein is also commercially available: Promega, United States Biochemical, New England Biolabs.
2. DNA helicase of interest, with activity defined in helicase units (*see* **Subheading 3.2.**).
3. SSB protein (*see* **Notes 1** and **3**). *E. coli* protein is commercially available: Promega and United States Biochemical. Purification of *E. coli* SSB protein has been described previously *(22)*.
4. DNA: 2 μg 5'-end-labeled linear pUC19 duplex DNA and 4 μg pUC19 scDNA, at stock concentration of 250 and 500 μ*M* (nucleotides), respectively (*see* **Note 2**).
5. 10X coupled reaction buffer: 250 m*M* Tris acetate, pH 7.5, 10 m*M* ATP, 10 m*M* PEP and 10 m*M* DTT. Store aliquots at –20°C.

3. Methods

3.1. Preparation of 5'-End-Labeled Linear DNA Substrate

For the best results, prepare the 5'-end-labeled linear duplex DNA substrate just prior to its use. The labeled DNA may be stored for approx 3 d at 4°C, but its quality rapidly degrades over time because of nicking of the DNA.

1. Digest and dephosphorylate 2 μg of pUC19 scDNA by incubation in 1X restriction buffer with 10 U of *Hind*III and 1 U shrimp alkaline phosphatase at 37°C for 1 h in a final volume of 20 μL.
2. Stop the reaction by heating at 70°C for 15 min to inactivate the enzymes. Cool to room temperature. Remove condensate from the lid of the tube by centrigugation in a microcentrifuge at 20,000*g* for 30 s. After this step the DNA can be stored at –20°C indefinitely.
3. To label the DNA with ^{32}P at the 5'-ends, add 20 μCi (γ-[^{32}P]) ATP and 10 U T4 polynucleotide kinase, and incubate at 37°C for 30 min.
4. Terminate the reaction by heating at 70°C for 15 min to inactivate T4 polynucleotide kinase. Cool to room temperature. Remove condensate from the lid of the tube by centrifugation in a microcentrifuge at 20,000*g* for 30 s.
5. Remove unincorporated label and exchange the restriction buffer for water using a MicroSpin G-25 column as per the supplier's instructions (*see* **Note 4**). Determine the volume of the recovered solution using a micropipettor, and calculate the concentration of DNA in μg/mL, assuming no loss of DNA (*see* **Note 5**). To calculate the approximate molar concentration of the DNA in μ*M* nucleotides,

divide the number of μg/mL by 0.325. This procedure typically yields a DNA solution having a concentration of between 200–250 μ*M* nucleotides.

3.2. DNA Helicase Assay

This section describes an agarose gel-based helicase assay to determine the specific activity of a DNA helicase (**Fig. 2**). In this general protocol, the DNA unwinding activity of the helicase can be assayed at any of the three different magnesium ion concentrations used in the coupled reaction (i.e., "high," 12 m*M* Mg-acetate; "intermediate", 6 m*M* Mg-acetate; and "low," 1 m*M* Mg-acetate) simply by altering the amount of 0.1 *M* Mg-acetate added to the reaction (*see* **Subheading 3.3.**). The substrate for unwinding is 5'-end-labeled linear pUC19 dsDNA having nearly blunt dsDNA ends, because this is the substrate for the helicase in the coupled reaction (*see* **Note 6**). After determining the rate of DNA unwinding for a known amount of DNA helicase, the specific activity of the preparation is calculated in terms of helicase units per milligram of protein. In this case, one helicase unit is defined as the amount of helicase needed to unwind 1 nmol basepairs in 1 min under each of the conditions described here ***(11)***. The specific activity of the DNA helicase should be measured at several different protein concentrations to obtain an accurate specific activity.

1. Aliquot 10 μL of freshly prepared stop solution into eight 1.5-mL microcentrifuge tubes. Label the tubes as "0 min," "15 s," "30 s," "60 s," "120 s," "300 s," "600 s," and "1200 s."
2. Prepare the reaction mixture (which will have a final volume of 200 μL after adding all the components) by adding 20 μL of 10X pairing buffer and 4 μL of pyruvate kinase to 100 μL of water in 1.5-mL microcentrifuge tubes. Mix the solution in the tube thoroughly using a vortex mixer.
3. Add 0.1 *M* Mg-acetate to achieve the desired final magnesium concentration: 24 μL for 12 m*M* Mg-acetate, 12 μL for 6 m*M* Mg-acetate, or 2 μL for 1 m*M* Mg-acetate. Test each of these magnesium ion concentrations individually to determine the specific activity of the helicase under each of the conditions to be used below in the coupled reaction. Mix the solution thoroughly with a vortex mixer.
4. To each tube, add SSB protein and 5'-end-labeled linear pUC19 dsDNA to a final concentration of 1 μ*M* and 10 μ*M*, respectively (*see* **Note 7**). Bring the volume of each reaction mixture up to 200 μL with water. Mix gently using a vortex mixer.
5. Incubate the reaction mixtures for 5 min at 37°C, then remove a 20-μL aliquot from the reaction, and add to the corresponding tube containing stop solution (i.e., "0 min"). Continue incubation of the remainder of the reaction mixture at 37°C. Mix the stopped aliquot thoroughly with vortex mixer and place on ice (*see* **Note 8**).
6. Initiate the unwinding reaction by adding the DNA helicase to a final concentration of 20 nM (calculate for a reaction volume of 180 μL) (*see* **Note 9**), mix gently, and continue incubation at 37°C. Take 20 μL aliquots at 15 s, 30 s, 60 s,

120 s, 300 s, 600 s, and 1200 s. Stop the reaction in each aliquot by adding to the appropriate tube containing the stop solution. Mix each aliquot thoroughly using vortex mixer and place on ice.

7. Deproteinize the DNA in each of the samples by incubating the aliquots at 37°C for 5 min.
8. Load 20 µL of each sample onto a 0.8% (w/v) agarose gel (13 × 26 × 0.5 cm) with 12 (1 × 0.15 cm) wells and run the gel at 1.5 V/cm for 12 h with 1X TAE as the running buffer. To visualize the DNA, dry the gel onto Whatman DE81 paper (*see* **Note 10**) and expose to either X-ray film or a phosphorimager screen (*see* **Note 11**). **Fig. 2A** is a representation of an autoradiogram for a typical time-course of RecQ-helicase-mediated unwinding of linear 5'-end-labeled dsDNA.
9. Determine the amount of radioactivity in the band corresponding to intact dsDNA at each time point, and calculate the percentage of dsDNA unwound by the helicase at each time-point (*see* **Note 11**).
10. Plot percent dsDNA unwound vs time as indicated in **Fig. 2B**. The resultant unwinding curve should be hyperbolic (*see* **Note 9**). Calculate the initial rate of unwinding in nM bp s^{-1} (n*M* helicase)$^{-1}$ from the slope of the initial linear portion of the curve using the following equation:

$$\text{initial rate} = \text{slope} \times (C/[100\%]) \times (1/X)$$

where C is the concentration of basepairs in n*M* (i.e., 5000 n*M*), and X is the concentration of DNA helicase in n*M*.

11. Calculate the specific activity of the helicase preparation in helicase units (mg of protein)$^{-1}$ for each magnesium concentration using the following equation:

$$\text{specific activity} = \text{initial rate} \times ([60\text{ s}]/[1\text{ min}]) \times ([1.0E + 6]/Y)$$

where Y is the molecular weight of the helicase. Use the calculated specific activity to determine the amount to DNA helicase to use in the coupled reaction (*see* **Subheading 3.4.**).

3.3. RecA Protein-Mediated D-Loop Formation—Partial Reaction

This section describes a protocol for optimizing the homologous pairing activity of an uncharacterized RecA-like protein by varying the magnesium ion concentration present in the buffer (*see* **Note 12**). The magnesium ion conditions are as follows: "high magnesium" (12 m*M* Mg-acetate), "intermediate magnesium" (6 m*M* Mg-acetate), and "low magnesium" (1 m*M* Mg-acetate with 10% PEG).

3.3.1. Preparation of the 5'-End-Labeled, Linear pUC19 ssDNA Substrate

Prepare this DNA substrate immediately before it is needed to ensure that the DNA added to the pairing reaction it is fully denatured.

1. Denature 2 µg of 5'-end-labeled linear pUC19 duplex DNA (*see* **Subheading 3.1.**) by heating at 95°C for 5 min.

2. Transfer the tube quickly to an ice-cold water bath and chill for 5 min. Remove condensate from the lid of the tube by centrifugation in microcentrifuge at 20,000*g* for 30 s at 4°C. Keep the denatured DNA on ice to prevent re-annealing of the ssDNA strands.

3.3.2. RecA Protein-Mediated D-Loop Assay

1. Aliquot 10 µL of freshly prepared stop solution into twelve 1.5-mL microcentrifuge tubes. Label sets of four tubes with "high," "intermediate," and "low." In each set, label a tube "0 min," "5 min," "15 min," and "30 min."
2. Prepare three reaction mixtures (which will have a final volume of 100 µL after adding all the components) by adding 10 µL of 10X pairing buffer and 2 µL of pyruvate kinase to 40 µL of water in 1.5-mL microcentrifuge tubes. Mix each tube well using a vortex mixer. Label one tube "high," the second "intermediate," and the last "low."
3. To the tube labeled "high," add 12 µL of 0.1 *M* Mg-acetate, to the tube labeled "intermediate," add 6 µL of 0.1 *M* Mg-acetate, and to the tube labeled "low," add 1 µL 0.1 *M* Mg-acetate. Also, add to the tube labeled "low," 25 µL of 40% PEG (*see* **Note 13**). Mix the solution in each tube well with a vortex mixer.
4. To each tube, add RecA protein and 5'-end labeled linear pUC19 heat-denatured ssDNA to a final concentration of 5 µ*M* and 10 µ*M*, respectively (*see* **Note 14**). Bring the volume of each reaction mixture up to 100 µL with water. Mix gently using a vortex mixer.
5. Incubate the reaction mixtures for 3 min at 37°C, then add SSB protein to a final concentration of 1 µ*M* (*see* **Note 14**) and return to incubation at 37°C. After an additional 2 min, remove a 20 µL aliquot from each reaction, and add to the corresponding tube containing stop solution (for example, "high, 0 min"). Continue incubation of the remainder of the reaction mixture at 37°C. Mix the stopped aliquots thoroughly with a vortex mixer and place on ice (*see* **Note 15**).
6. Initiate the homologous pairing reaction by adding pUC19 scDNA to a final concentration of 20 µ*M* (calculate for a reaction volume of 80 µL), mix gently, and continue incubation at 37°C. Take 20-µL aliquots at 5, 15, and 30 min. Stop each pairing reaction by adding each aliquot to the appropriate tube containing the stop solution. Mix each aliquot thoroughly using a vortex mixer and place on ice.
7. Deproteinize the DNA in each of the samples by incubating the aliquots at 37°C for 5 min.
8. Load 20 µL of each sample onto a 1% (w/v) agarose gel (13 × 26 × 0.5 cm) with 12 (1 × 0.15 cm) wells and run the gel at 1.5 V/cm for 12 h with 1X TAE as the running buffer. To visualize the DNA, dry the gel onto Whatman DE81 paper (*see* **Note 10**) and expose to either X-ray film or a phosphorimager screen. **Fig. 1B** is a representation of an autoradiogram for a typical time-course of the D-loop reaction mediated by RecA protein and 5'-end-labeled linear ssDNA.

3.4. Coupling the Activities of RecA Protein and a DNA Helicase to Form D Loops—Coupled Reaction

The protocol below describes a procedure to couple the activities of a DNA helicase and RecA protein to produce D loops. To optimize the reaction, D-loop formation is assayed under three different magnesium concentrations (*see* **Note 16**): "high magnesium" (12 m*M* Mg-acetate), "intermediate magnesium" (6 m*M* Mg-acetate) and "low magnesium" (1 m*M* Mg-acetate with 10% PEG).

1. Aliquot 10 µL of freshly prepared stop solution into twelve 1.5-mL microcentrifuge tubes. Label sets of four tubes with "high," "intermediate," and "low." In each set, label a tube "0 min," "5 min," "15 min," and "30 min."
2. Prepare three reaction mixtures by adding 10 µL of 10X coupled reaction buffer and 2 µL of pyruvate kinase to 40 µL of water in 1.5-mL microcentrifuge tubes. Label one tube "high," the second "intermediate," and the last "low."
3. To the tube labeled "high,add 12 µL of 0.1 *M* Mg-acetate, to the tube labeled "intermediate" add 6 µL of 0.1 *M* Mg-acetate, and to the tube labeled "low," add 1 µL 0.1 *M* Mg-acetate. Also, add to the tube labeled "low," 25 µL of 40% PEG (*see* **Note 13**). Mix the solution in each tube well with a vortex mixer.
4. To each tube, add RecA protein and SSB protein to a final concentration of 5 µ*M* and 1 µ*M*, respectively. Also add pUC19 scDNA and 5'-end-labeled linear pUC19 duplex DNA to a final concentration of 20 µ*M* and 10 µ*M*, respectively (*see* **Note 14**). Bring the volume of each reaction mixture up to 100 µL with water. Mix gently using a vortex mixer.
5. Incubate the reaction mixtures for 5 min at 37°C. Remove a 20 µL aliquot from each reaction, and add to the corresponding tube containing stop solution (for example "high, 0 min"). Continue incubation of the remainder of the reaction mixture at 37°C. Mix the stopped aliquots thoroughly with a vortex mixer and place on ice (*see* **Note 15**).
6. Start the reaction by adding 20 helicase U of the DNA helicase (*see* **Note 17**), mix gently, and continue incubation at 37°C. Take additional aliquots at 5, 15, and 30 min and stop the reaction by adding each aliquot to the appropriate tube containing the stop solution. Mix each aliquots thoroughly using a vortex mixer and place on ice.
7. Deproteinize the DNA in each of the samples by incubating the aliquots at 37°C for 5 min.
8. Load 20 µL of each sample onto a 1% (w/v) agarose gel (13 × 26 × 0.5 cm) with 12 (1 × 0.15 cm) wells and run the gel at 1.5 V/cm for 12 h with 1X TAE as the running buffer. To visualize the DNA, dry the gel onto Whatman DE81 paper (*see* **Note 10**) and expose to either X-ray film or a phosphorimager screen. **Fig. 1C** is a representation of an autoradiogram for a typical time-course of D-loop formation mediated by RecQ helicase and *E. coli* RecA protein.

4. Notes

1. SSB protein is present in these reactions to trap the ssDNA produced by a DNA helicase so that this ssDNA does not reanneal after passage of the helicase ***(11,25–27)***. Because of differences in helicases from species to species, it is desirable, where possible, to use a DNA helicase and an SSB protein from the same or related species.
2. The molar nucleotide concentration of duplex DNA can be determined spectrophotometrically using the extinction coefficient of 6500 $M^{-1}cm^{-1}$ at a wavelength of 260 nm ***(16)***.
3. SSB protein is present in these reactions to remove secondary structure in ssDNA, which inhibits RecA protein function ***(28)***. Because of differences in recombination proteins from species to species, it is desirable, where possible, to use RecA-like and SSB proteins from the same or related species.
4. Exchange of unincorporated label and buffer can also be achieved by precipitating the DNA with ethanol twice ***(21)***. In this case, determine the DNA concentration spectrophotometrically using the extinction coefficient of 6500 $M^{-1}cm^{-1}$ at a wavelength of 260 nm. To confirm that all the ATP has been removed, calculate the ratio of the absorbance at 260 nm to the absorbance at 280 nm. The value for this ratio should be 1.8–1.9; a higher value than this standard is indicative of contamination by ATP.
5. In our hands, DNA loss after this step is <10%, which is acceptable for these experiments. It is not unusual to gain between 2–10 µL of volume from the spin column; therefore, it is important to determine the final volume of the DNA solution. Please note that this treatment of the DNA imparts ultraviolet light-absorbent material to the DNA solution, so the DNA concentration cannot be accurately determined spectrophotometrically. In most cases, the absorbent material can be washed out of the column by five successive washes with water (totaling approx 5 column volumes). To wash the column once, centrifuge column as per the suppliers' instructions to remove the original buffer, resuspend the matrix in 400 µL of water, then centrifuge as before. After washing in this manner five times, follow the suppliers' instructions for exchange of unincorporated label and buffer in the DNA sample. The concentration of the recovered DNA can then be determined spectrophotometrically as described in **Note 3**.
6. If the DNA helicase of interest is incapable of initiating unwinding from flush dsDNA ends, then the DNA substrate must be modified to contain the appropriate initiation site for the DNA helicase. In most cases, the addition of short (>100 nt) ssDNA tails are sufficient for most DNA helicases, but the polarity of the tail must be determined for each helicase. Linear dsDNA can be resected to contain ssDNA tails by use of a dsDNA-dependent exonuclease ***(29,30)***; for example, *E. coli* Exonuclease III will generate 5'-tails, and λ Exonuclease will produce 3'-tails. Exonuclease III and λ Exonuclease are available from United States Biochemical. Keep in mind that if tailed DNA substrates are used in the coupled reaction, the ssDNA regions must be heterologous to the scDNA recipient. Otherwise, RecA protein, alone, will pair the tailed dsDNA with the scDNA recipient without the need for a DNA helicase.

7. The concentration of SSB protein used here produces optimal unwinding for both RecQ helicase and RecBCD enzyme. For an uncharacterized helicase, the concentration of SSB protein should be varied to achieve maximal unwinding activity. Keep in mind that the ssDNA binding stoichiometry of *E. coli* SSB protein is dependent on the ionic conditions of the solution: at magnesium ion conditions greater than 1–2 m*M* (at 37°C), the site size is approx 15 nucleotides per SSB monomer ***(31)***.
8. The aliquots are kept on ice to prevent the re-annealing of the individual ssDNA strands.
9. For the best results, the concentration of DNA helicase in the reaction should be limiting relative to the duplex DNA substrate so that the initial rate of unwinding can be accurately measured. For an uncharacterized DNA helicase, vary the DNA helicase concentration over a 10- to 100-fold range to achieve a hyperbolic unwinding curve with an obvious linear initial slope (*see* **Fig. 2B**).
10. The agarose gel is dried onto Whatman DE81 paper to minimize DNA loss. Gel drying time can be reduced by flattening the gel prior to drying. To flatten the gel, place the gel on Whatman DE81 paper cut to the size of the gel, which is atop several sheets of filter paper and lab matting. Cover the gel with a single layer of plastic wrap. Place a rigid sheet of Lucite or a wooden board over the gel, followed by a 15–20-lb. weight. Dry the gel onto the Whatman DE81 paper once the gel is between 1–2 mm in thickness.
11. An alternative method to visualize the DNA is to stain the wet gel with 0.5 μg/mL ethidium bromide in water for 1 h, destain with water for 1 h, and illuminate the gel using an ultraviolet light box. In this case, the amount of radioactivity in each band can be determined by excising the individual bands and analyzing each with a scintillation counter.
12. Each buffer condition provided here supports homologous pairing and DNA strand exchange by *E. coli* RecA protein. The magnesium requirements for *E. coli* RecA protein are ≈4 m*M* magnesium ion in dilute solution ***(32)***, or as low as 1 m*M* magnesium ion in a buffer supplemented with a volume exclusion agent, such as the PEG used here ***(33)***.
13. The 40% PEG solution is extremely viscous, so the solution must be drawn up into the micropipet slowly to measure the proper volume of PEG. Pipet the PEG solution into the buffer slowly, taking care to expel all of the solution into the microcentrifuge tube. Mix well by pipeting the buffer and PEG mixture up and down several times until a homogenous mixture is obtained.
14. The concentrations of RecA protein and SSB protein used here produce optimal results when using *E. coli* RecA protein. For an uncharacterized RecA-like protein, vary the concentration of each protein to achieve maximal pairing activity. Keep in mind that a saturating amount of RecA protein is achieved at a stoichiometry of 1 RecA monomer to 3 nucleotides of ssDNA ***(34)***. In addition, the ssDNA binding stoichiometry of *E. coli* SSB protein is dependent on the ionic conditions of the solution: at magnesium ion conditions greater than 1–2 m*M* (at 37°C), the site size is approx 15 nucleotides per SSB monomer ***(31)***. It is

also important to note excessive amounts of SSB protein or addition of SSB protein to ssDNA before RecA protein will inhibit RecA protein, because these two proteins compete for binding to ssDNA ***(28)***. Also, though an excess of *E. coli* RecA protein does not inhibit DNA strand exchange, excessive amounts of *Saccharomyces cerevisiae* Rad51 protein inhibit DNA strand exchange by binding to the dsDNA ***(35)***.

15. Each time-point is placed on ice to minimize thermal dissociation of the D loops formed by RecA protein.
16. RecA protein and the DNA helicase may not display optimal activity at the same concentration of magnesium, so it is important to test a range of magnesium concentrations ***(16,17)***. As stated above (*see* **Note 12**), *E. coli* RecA protein is capable of pairing under all the conditions presented here.
17. Helicase units are defined and determined in **Subheading 3.2.** In our hands, 20 U of helicase activity produce sufficient ssDNA for use as a substrate for *E. coli* RecA protein. This amount of helicase corresponds to approx 2.5 n*M* RecBCD enzyme ***(17)*** or 100 n*M E. coli* RecQ helicase ***(16)***. However, the amount of DNA helicase needed for optimal coupled pairing must be determined empirically for each DNA helicase. To optimize the reaction, DNA helicase concentration should be varied over a 10- to 100-fold range. Keep in mind that the optimum may be very sharp, because the DNA helicase being tested may also be capable of disrupting the joint molecules formed by RecA protein, as is the case for the *E. coli* RecQ helicase ***(16)***.

References

1. Friedberg, E. C., Walker, G. C., and Siede, W. (1995) *DNA Repair and Mutagenesis*, ASM, Washington, DC.
2. Kowalczykowski, S. C., Dixon, D. A., Eggleston, A. K., et al. (1994) Biochemistry of homologous recombination in *Escherichia coli. Microbiol. Rev.* **58,** 401–465.
3. Kowalczykowski, S. C. and Eggleston, A. K. (1994) Homologous pairing and DNA strand-exchange proteins. *Annu. Rev. Biochem.* **63,** 991–1043.
4. Roca, A. I. and Cox, M. M. (1997) RecA protein: structure, function, and role in recombinational DNA repair. *Prog. Nucleic Acid Res. Mol. Biol.* **56,** 129–223.
5. Bianco, P. R., Tracy, R. B., and Kowalczykowski, S. C. (1998) DNA strand exchange proteins: a biochemical and physical comparison. *Front Biosci.* **3,** D570-D603.
6. Ogawa, T., Shinohara, A., Nabetani, A., et al. (1993) RecA-like recombination proteins in eukaryotes: Function and structures of *RAD51* genes. *Cold Spring Harbor Symp. Quant. Biol.* **58,** 567–576.
7. Sargentini, N. J. and Smith, K. C. (1986) Characterization and quantitation of DNA strand breaks requiring recA- dependent repair in X-irradiated Escherichia coli. *Radiat. Res.* **105,** 180–186.
8. Clark, A. J. and Margulies, A. D. (1965) Isolation and characterization of recombination-deficient mutants of *Escherichia coli* K12. *Proc. Natl. Acad. Sci. USA* **53,** 451–459.

9. Cox, M. M. and Lehman, I. R. (1982) RecA protein-promoted DNA strand exchange: Stable complexes of recA protein and single-stranded DNA formed in the presence of ATP and single-stranded DNA binding protein. *J. Biol. Chem.* **257,** 8523–8532.
10. Taylor, A. and Smith, G. R. (1980) Unwinding and rewinding of DNA by the recBC enzyme. *Cell.* **22,** 447–457.
11. Roman, L. J. and Kowalczykowski, S. C. (1989) Characterization of the helicase activity of the *Escherichia coli* RecBCD enzyme using a novel helicase assay. *Biochemistry* **28,** 2863–2873.
12. Eggleston, A. K. and West, S. C. (1997) Recombination initiation: easy as A, B, C, D... chi? *Curr. Biol.* **7,** R745–749.
13. Arnold, D. A. and Kowalczykowski, S. C. (1999) RecBCD nuclease/helicase, in *Encyclopedia of Life Sciences*, Nature Publishing Group, London, http://www.els.net.
14. Nakayama, H., Nakayama, K., Nakayama, R., et al. (1984) Isolation and genetic characterization of a thymineless death-resistant mutant of *Escherichi coli* K-12: Identification of a new mutation (*recQ1*) that blocks the recF recombination pathway. *Mol. Gen. Genet.* **195,** 474–480.
15. Lanzov, V., Stepanova, I., and Vinogradskaja, G. (1991) Genetic control of recombination exchange frequency in *Escherichia coli* K-12. *Biochimie* **73,** 305–312.
16. Harmon, F. G. and Kowalczykowski, S. C. (1998) RecQ helicase, in concert with RecA and SSB proteins, initiates and disrupts DNA recombination. *Genes Dev.* **12,** 1134–1144.
17. Roman, L. J., Dixon, D. A., and Kowalczykowski, S. C. (1991) RecBCD-dependent joint molecule formation promoted by the *Escherichia coli* RecA and SSB proteins. *Proc. Natl. Acad. Sci. USA* **88,** 3367–3371.
18. Dixon, D. A. and Kowalczykowski, S. C. (1991) Homologous pairing in vitro stimulated by the recombination hotspot, Chi. *Cell* **66,** 361–371.
19. West, S. C., Cassuto, E., and Howard-Flanders, P. (1981) RecA protein promotes homologous-pairing and strand-exchange reactions between duplex DNA molecules. *Proc. Natl. Acad. Sci. USA* **78,** 2100–2104.
20. Matson, S. W., Tabor, S., and Richardson, C. C. (1983) The gene 4 protein of bacteriophage T7 Characterization of helicase activity. *J. Biol. Chem.* **258,** 14,017–14,024.
21. Sambrook, J., Fritsch, E. F., and Maniatis, T. (eds.) (1989) *Molecular Cloning: A Laboratory Manual, Second Edition*, Cold Spring Harbor Laboratory Press, Cold Spring Harbor, NY.
22. LeBowitz, J. (1985) Biochemical mechanism of strand initiation in bacteriophage lambda DNA replication. Ph.D. thesis, Johns Hopkins University, Baltimore, MD.
23. Cox, M. M., McEntee, K., and Lehman, I. R. (1981) A simple and rapid procedure for the large scale purification of the recA protein of *Escherichia coli*. *J. Biol. Chem.* **256,** 4676–4678.
24. Griffith, J. and Shores, C. G. (1985) RecA protein rapidly crystallizes in the presence of spermidine: a valuable step in its purification and physical characterization. *Biochemistry* **24,** 158–162.

25. Matson, S. W. and George, J. W. (1987) DNA helicase II of *Escherichia coli*. Characterization of the single-stranded DNA-dependent NTPase and helicase activities. *J. Biol. Chem.* **262,** 2066–2076.
26. Kowalczykowski, S. C. and Roman, L. J. (1990) Reconstitution of homologous pairing activity dependent upon the combined activities of purified *E. coli* RecA, RecBCD, and SSB proteins, in *Molecular Mechanisms in DNA Replication and Recombination* (Richardson, C. C. and Lehman, I. R., eds.), Wiley-Liss, New York, pp. 357–373.
27. Umezu, K. and Nakayama, H. (1993) RecQ DNA helicase of *Escherichia coli*. Characterization of the helix-unwinding activity with emphasis on the effect of single-stranded DNA-binding protein. *J. Mol. Biol.* **230,** 1145–1150.
28. Kowalczykowski, S. C. and Krupp, R. A. (1987) Effects of the *Escherichia coli* SSB protein on the single-stranded DNA-dependent ATPase activity of *Escherichia coli* RecA protein: evidence that SSB protein facilitates the binding of RecA protein to regions of secondary structure within single-stranded DNA. *J. Mol. Biol.* **193,** 97–113.
29. Register, J. C., III and Griffith, J. (1985) The direction of RecA protein assembly onto single strand DNA is the same as the direction of strand assimilation during strand exchange. *J. Biol. Chem.* **260,** 12,308–12,312.
30. Anderson, D. G. and Kowalczykowski, S. C. (1997) The recombination hot spot c is a regulatory element that switches the polarity of DNA degradation by the RecBCD enzyme. *Genes Dev.* **11,** 571–581.
31. Bujalowski, W. and Lohman, T. M. (1986) Escherichia coli single-strand binding protein forms multiple, distinct complexes with single-stranded DNA. *Biochemistry* **25,** 7799–7802.
32. Lavery, P. E. and Kowalczykowski, S. C. (1990) Properties of recA441 protein-catalyzed DNA strand exchange can be attributed to an enhanced ability to compete with SSB protein. *J. Biol. Chem.* **265,** 4004–4010.
33. Lavery, P. E. and Kowalczykowski, S. C. (1992) Enhancement of recA protein-promoted DNA strand exchange activity by volume-occupying agents. *J. Biol. Chem.* **267,** 9307–9314.
34. Cox, M. M. and Lehman, I. R. (1981) Directionality and polarity in recA protein-promoted branch migration. *Proc. Natl. Acad. Sci. USA* **78,** 6018–6022.
35. Sung, P. (1994) Catalysis of ATP-dependent homologous DNA pairing and strand exchange by yeast RAD51 protein. *Science* **265,** 1241–1243.

8

Detection and Quantitation of RecBCD Enzyme (Exonuclease V) Activity

Douglas A. Julin

1. Introduction

1.1. Introduction to RecBCD Enzyme and Its Activities

The RecBCD enzyme serves two functions in the bacterial cell: it is a nuclease that destroys linear double-stranded DNA (dsDNA), and a DNA helicase that generates single-stranded DNA (ssDNA) used by RecA protein to initiate homologous recombination ***(1–3)***. A specific DNA sequence called Chi (5'-GCTGGTGG) is a signal that regulates these two functions ***(1,3)***. An encounter with Chi by RecBCD during its reaction with dsDNA leads to suppression of the nuclease activity and reveals the recombination-initiating function of RecBCD ***(1,3)***. The enzyme in *Escherichia coli* and other bacteria consists of three protein subunits encoded by the *recB*, *recC*, and *recD* genes, whereas some bacteria (e.g., *Bacillus subtilis*, *Lactococcus lactis*) produce a two-subunit enzyme [AddAB and RexAB, respectively ***(4,5)***]. The genes encoding these latter enzymes complement an *E. coli recBCD* deletion mutation in vivo ***(5,6)***, and the purified AddAB enzyme has similar catalytic activity to the more extensively studied *E. coli* enzyme ***(7)***.

RecBCD is a multifunctional enzyme in vitro, befitting its multiple biological functions. The enzyme is considered to have at least four catalytic activities: adenosine triphosphate (ATP)-dependent exonuclease on linear ss- or dsDNA, ATP-stimulated endonuclease on circular ssDNA, ss- or dsDNA-dependent ATP hydrolysis, and ATP-dependent DNA helicase ***(3,8,9)***. In addition, the interaction with Chi in dsDNA can be assayed in vitro ***(10,11)***.

ATP-dependent degradation of linear dsDNA was the first activity associated with RecBCD enzyme and led to its designation as Exonuclease V ***(9,12–14)***.

From: *Methods in Molecular Biology, vol. 152: DNA Repair Protocols: Prokaryotic Systems*
Edited by: P. Vaughan © Humana Press Inc., Totowa, NJ

This reaction requires a linear dsDNA substrate, magnesium ion, and ATP or other hydrolyzable nucleoside triphosphate ***(14,15)***. The enzyme has no detectable nuclease activity with circular dsDNA, whether supercoiled or nicked ***(14,15)***.

RecBCD also degrades linear ssDNA, such as denatured linearized plasmid DNA. This ss-exonuclease reaction also requires magnesium ion and ATP when it is assayed by the production of small acid-soluble products as described below ***(15)***. However, RecBCD cleaves circular ssDNA such as that of bacteriophage M13 endonucleolytically in a reaction that is detectable without ATP but is stimulated *ca.* seven-fold when ATP is included ***(15)***. The subsequent degradation of the linearized M13 DNA is not detected without ATP ***(15)***. Nuclease activity in the absence of ATP can be detected on linear ssDNA using end-labeled oligomeric DNA substrates [15–25 nucleotides (nt) ***(16)***]. The reactions with these small ssDNA substrates are nonetheless approx 3000-fold faster with 0.5 mM ATP than without ***(16)***.

RecBCD hydrolyzes ATP in the presence of all DNA substrates on which it is a nuclease (linear dsDNA and linear or circular ssDNA, but not circular dsDNA), and magnesium ion or other divalent cation ***(14,15,17)***. There is no detectable ATP hydrolysis in the absence of DNA ***(15)***.

The helicase activity of RecBCD was discovered as investigators sought to understand the unusual requirement by a nuclease for ATP ***(17–19)***. The nuclease activity can be suppressed by altering the reaction conditions, particularly by manipulating the concentration and type of divalent metal ion ***(17)***, permitting the detection and study of DNA unwinding. The DNA substrate is the same as that for the ds-exonuclease activity: linear, but not circular, dsDNA ***(20)***, and, as with other helicases ***(21)*** the unwinding reaction requires ATP. The Mg:ATP ratio is the key determinant of the nuclease cleavage frequency relative to the DNA unwinding rate. In general, the nuclease reaction requires that the Mg^{2+} concentration exceed that of ATP ***(17,22)***, as both Mg·ATP and additional free Mg^{2+} are presumably required for this activity. The nuclease activity is reduced in reaction mixtures where the Mg^{2+} concentration is less than that of ATP, or when Ca^{2+} is also included, and RecBCD unwinds the DNA under either condition. Thus, DNA unwound by RecBCD in reaction mixtures containing 1 m*M* $MgCl_2$, 1 m*M* $CaCl_2$, and 5 m*M* ATP has been visualized by electron microscopy ***(19,23)***. *E. coli* single-stranded DNA binding protein (SSB) is generally included to bind the unwound DNA strands and prevent reannealing. SSB protein reduces, but does not eliminate the production of small nuclease products from dsDNA ***(24)***. On the other hand, SSB protein completely blocks both ssDNA-dependent ATP hydrolysis and nuclease cleavage by RecBCD ***(24,25)***. The reactions with dsDNA (helicase and helicase/nuclease) are rapid [approx 200–1000 basepairs unwound/s ***(19,26)***] and highly processive [approx 30,000 basepairs traveled per binding event ***(27)***].

The above discussion indicates that the properties and products of the RecBCD-catalyzed reactions depend strongly on the DNA substrate structure and the reaction conditions. Moreover, the designation of RecBCD as either an endonuclease or exonuclease has depended on the substrate and assay method used. The ATP-dependent ds-exonuclease reaction catalyzed by RecBCD is thought to involve ATP-dependent translocation along and unwinding of the dsDNA, with concomitant degradation of the partially unwound strands produced by the helicase ***(23,24)***. The nuclease reaction products are small ss-oligonucleotides, but not mononucleotides ***(15)***, and thus the "exonuclease" reaction is in fact an endonuclease by the strict definition of endo- vs exonuclease ***(28)***.

The ds-exonuclease activity of RecBCD from *E. coli* is inhibited, but unwinding continues, when the enzyme encounters a specific sequence called Chi (5'-GCTGGTGG) in the DNA ***(11,29)***. This "Chi activity" can be detected in vitro by the production of a specific, 5'-end-labeled ssDNA fragment extending from the Chi site to the end of the strand containing Chi ***(10,11)***. Chi activity has been detected in RecBCD enzymes from a variety of bacteria ***(30)***, and sequences have been identified in other organisms that function similarly to Chi in *E. coli* ***(5,31)***.

1.2. Diagnostic Assay for RecBCD Enzyme Activity

The ATP-dependent exonuclease reaction on linear dsDNA is a convenient way to detect and assay RecBCD enzyme activity. The assay is sensitive and background activity from other nucleases can be determined by performing reactions without added ATP, allowing the enzyme activity to be measured in crude cell lysates. The most straightforward assay uses a tritium-labeled dsDNA substrate and quantitates the production of small oligonucleotide fragments [< approx 15–20 nt ***(32)***] that are soluble in cold acidic solution.

2. Materials

2.1. [^{3}H]DNA Preparation

1. M9CA medium: 10X M9 salts: Na_2HPO_4 (60 g/L), KH_2PO_4 (30 g/L), NaCl (5 g/L), NH_4Cl (10 g/L), adjusted to pH 7.4 with phosphoric acid, HCl, or 5*M* NaOH. Dilute 10 mL of 10X M9 salts to 100 mL with water in a 0.5-liter flask and autoclave. After the autoclaved liquid has cooled, add sterile: Casamino acids (1 mL of 20% (w:v, i.e., 20 g/100 mL), autoclaved), glucose (1 mL of 20% (w:v), autoclaved), $CaCl_2$ (0.01 mL of 1 *M*, filter-sterilized), $MgSO_4$ (0.2 mL of 1 *M*, filter-sterilized). Mix to redissolve the $CaPO_4$ precipitate that forms.
2. [*methyl*-^{3}H]thymidine, from Amersham Pharmacia Biotech, #TRK686, 1 mCi/mL, 70–86 Ci/mmol.
3. Uridine (Sigma).
4. 50 m*M* Tris-HCl, pH 8.0, 50 m*M* ethylenediaminetetraacetic acid (EDTA).

5. Egg white lysozyme, Grade I, from Sigma. Prepare a fresh 10 mg/mL solution in water and discard unused portion.
6. 0.5% (w:v) SDS, 50 m*M* Tris-HCl, pH 7.5, 0.4 *M* EDTA.
7. Proteinase K (Sigma).
8. Equilibrated phenol/chloroform (1:1, v:v).
9. 3 *M* sodium acetate and 100% ethanol.
10. 50 m*M* Tris-HCl, pH 7.5, 1 m*M* EDTA.
11. RNase A (*see* **Note 1**).
12. QIAGEN Plasmid Kit (QIAGEN Inc., Valencia, CA).
13. TE8.0: 10 m*M* Tris-HCl, 1 m*M* EDTA, pH 8.0.
14. Restriction endonuclease and 10X reaction buffer.
15. Materials for agarose gel electrophoresis.

2.2. Cell-Free Extract Preparation

1. 10 m*M* Tris-HCl, pH 7.5, 20% sucrose (w:v).
2. 0.5 *M* β-mercaptoethanol: mix 35 μL of pure β-mercaptoethanol (14.4 *M*) with 965 μL water, and store in the refrigerator.
3. 5 *M* NaCl.

2.3. Stock Solutions for RecBCD Enzyme Assay

1. 4X Reaction Buffer: 200 m*M* Tris-HCl, pH 8.5, 40 m*M* $MgCl_2$, 0.8 mg/mL bovine serum albumin (BSA), 2.68 m*M* dithiothreitol (DTT).
2. RecBCD Dilution Buffer: 10 m*M* potassium phosphate, pH 6.7, 0.4 mg/mL BSA, 0.1 m*M* EDTA, 0.1 m*M* DTT, 20% (v:v) glycerol.

Tris-HCl, pH 8.5, potassium phosphate, pH 6.7, and $MgCl_2$ stocks (1 *M* each), and Na·EDTA (0.5 *M*, adjusted to pH 8.0 with NaOH), should be autoclaved. BSA (10 mg/mL; *see* **Note 2**) and DTT (0.1 *M*) can be filter-sterilized. The 4X reaction buffer, RecBCD dilution buffer, BSA, and DTT stocks should be stored at –20°C in small aliquots.

3. ATP (1 m*M*). ATP can be purchased as a premade solution (e.g., Amersham Pharmacia Biotech) or as the disodium salt from Sigma. A 100 m*M* stock solution is prepared from the latter by dissolving 0.55 g of Na_2·ATP (551.1 g/mol) in 8 mL of water. Adjust the pH to 7–8.0 by adding 5 *M* NaOH and then add water to 10 mL total volume. Dilute with water to a working 1 m*M* stock. The solutions should be stored in small aliquots frozen at –20°C (*see* **Note 3**).
4. 10% Trichloroacetic acid (TCA). Prepare a 100% (w:v) TCA solution by adding 227 mL of water to a 500 g bottle of solid TCA (**ref.** ***(33)***, p. B.13). Prepare the 10% TCA quench solution by diluting 10 mL of 100% (w/v) TCA to 100 mL with water. Store both in the refrigerator (*see* **Note 4**).
5. Calf thymus DNA, 0.5 mg/mL. Dissolve 50 mg of calf thymus DNA (sodium salt; Sigma) in 100 mL of TE 8.0 with stirring and gentle heating (*see* **Note 5**).
6. Liquid scintillation cocktail that is miscible with aqueous sample (e.g., Bio-Safe II, Research Products International Corp., Mount Prospect, IL).

3. Methods

3.1. Tritium-Labeled DNA Substrates

E. coli chromosomal DNA or plasmid DNA can be internally labeled with tritium by growth in minimal medium containing [^{3}H]thymidine. *E. coli* DNA can be prepared easily in large amounts, wheras linear plasmid DNA has the advantage that the substrate structure is well defined and homogeneous. The disadvantage of the latter is that the circular plasmid must be cleaved with a restriction endonuclease before use.

3.1.1. E. Coli Chromosomal [^{3}H]DNA

Many *E. coli* strains can be tritium-labeled by growth in minimal medium such as M9CA. We have prepared chromosomal DNA from *E. coli* strain K37 (ATCC #33626).

1. Grow an overnight culture of the chosen strain in 1–2 mL of sterile M9CA medium with any other required nutrients (*see* **Note 6**). Inoculate the culture with a single colony picked from a plate, or gouge some cells from a frozen culture (–80°C) with a sterile needle.
2. The next morning, pour the saturated overnight culture into 100 mL of fresh M9CA medium and shake at 37°C. Measure the OD_{600} of the culture every 30–45 min. When the culture reaches OD_{600} = 0.1–0.2 (i.e., just entering log phase), add 0.5–1 mL of [*methyl*-^{3}H]thymidine (1 mCi/mL, *see* **Note 7**) and 20 mg of uridine (*see* **Note 8**).
3. Allow the culture to grow to saturation (8–12 h) at 37°C with shaking.
4. Harvest the cells by centrifugation at 5000*g* for 10 min in a JA14 rotor (Beckman) or GSA rotor (Sorvall). Pour off the culture supernatant and dispose as radioactive waste. Store the pellet at –80°C overnight, or proceed (*see* **Note 9**).
5. Resuspend the cell pellet in 10 mL of 50 m*M* Tris-HCl, pH 8.0, 50 m*M* EDTA. Transfer to a 50-mL plastic centrifuge tube.
6. Add 10 mg of lysozyme and put the tube on ice for 45 min.
7. Add 2 mL of 0.5% (w:v) SDS, 50 m*M* Tris-HCl, pH 7.5, and 0.4 *M* EDTA. Add proteinase K to a final concentration of 1 mg/mL. Mix well and incubate at 50°C for 1 h.
8. Shear the DNA that is released into solution by sonication or by forcing it through a syringe and needle. Start with a 16-gage needle, and force the solution through successively narrower needles (i.e., 18, 20, 23, 25 gage) (*see* **Note 10**).
9. Add an equal volume of equilibrated phenol/chloroform (1:1, v:v). Mix well.
10. Centrifuge for 5–15 min at 7000*g* to separate the layers. Transfer the upper aqueous phase to a new tube, leaving behind as much as possible of the white material at the aqueous/organic interface.
11. Add 3 mL of 3 *M* sodium acetate and 2 vol of ethanol. Mix well. Remove the precipitated DNA/RNA by centrifugation (30 min, 10,000*g*). Redissolve in 25 mL of 50 m*M* Tris-HCl, pH 7.5, 1 m*M* EDTA, at room temperature.
12. Add RNase A to 0.2 mg/mL and incubate at 37°C for 30 min.

13. Add an equal volume of phenol/chloroform. Mix and centrifuge. Transfer the aqueous phase to a new tube.
14. Add 2 vol of ethanol. Sediment the DNA pellet by centrifugation (30 min, 10,000*g*) and dissolve it in 2–5 mL of TE 8.0.
15. Measure the absorbance at 260 and 280 nm to determine the DNA concentration (*see* **Note 11**). The DNA concentration in µg/mL = $A_{260} \times 50$, for a 1-cm path-length cell. The molar concentration in terms of nucleotide residues is given by: *M* nt residues = µg DNA/mL × 1 µg/1000 mg × 1 mol nt / 324.5 g. The concentration should be about 1 m*M* nt residues.
16. Determine the specific radioactivity (cpm/nmol) by counting several small samples (1 µL each, mixed with 199 µL water, and 3 mL of liquid scintillation fluid) in the [³H] channel of the scintillation counter:

$$\text{specific radioactivity (cpm/nmol nt)} = \frac{\text{cpm/µL}}{\text{[DNA] (mol/L)}} \times 10^6\ \text{µL/L} \times 1\ \text{mol}/10^9\ \text{nmol}$$

The specific radioactivity should be 20,000–50,000 cpm/nmol nt residues (*see* **Note 12**).

3.1.2. Preparing Tritium-Labeled Plasmid DNA

1. Grow an overnight culture of cells containing the plasmid in 5 mL of M9CA medium containing antibiotic and other nutrients as required for the particular strain (*see* **Notes 6** and **13**).
2. The next day, pour the saturated overnight culture into 0.5 L of fresh M9CA medium + other nutrients + antibiotic and shake at 37°C. Monitor the growth by OD_{600}, and add [³H]thymidine (2.5 mL) and uridine (100 mg) as above (**step 2**, **Subheading 3.1.1.**). Harvest the cells after 8–12 h at 37°C by sedimentation, as above (**step 4**, **Subheading 3.1.1.**).
3. The plasmid DNA can be isolated quickly using a number of commercial products. We have used the QIAGEN Plasmid Kit (QIAGEN). Alternatively, the plasmid DNA can be isolated from cleared lysates by centrifugation on cesium chloride/ethidium bromide gradients (*ref.* *(33)*, pp. 1.38–1.45).
4. The DNA concentration and specific radioactivity should be determined by absorbance measurement and liquid scintillation counting, as above (**steps 15** and **16**, **Subheading 3.1.1.**, and *see* **Note 12**). The [³H]plasmid DNA, recovered in 1 mL TE8.0, is generally about 500 µM nt and 20,000–100,000 cpm/nmol nt.
5. Treat the circular plasmid with a restriction endonuclease to linearize it and purify by phenol/chloroform extraction and ethanol precipitation (*see* **Note 14**). Dissolve the DNA pellet in TE8.0. Count the radioactivity in a small sample of the cleaved [³H]DNA (approx 1 µL) to determine the DNA concentration (*see* **Note 12**). The DNA concentration is given by:

$$\text{[DNA] (µ}M\text{ nt)} = \frac{\text{cpm}}{\text{cpm/nmol} \times (1\ \text{µL})} \times 10^3$$

The DNA concentration in terms of duplex ends, the actual substrate for RecBCD, is given by:

$$\mu M \text{ DNA ends} = [\text{DNA}]\ (\mu\text{M nt})/(\#\ \text{bp/molecule})$$

3.2. RecBCD Enzyme Preparation

RecBCD enzyme is present at very low levels in wild-type cells [approx 10 enzyme molecules/cell ***(19)***]. Nonetheless, the nuclease activity has been measured in crude lysates of these cells, relying on the low background of other ATP-dependent nucleases. The *recB*, *recC*, and *recD* genes of *E. coli* have been cloned on an 18.5 kb *Bam*HI fragment in several different vectors including pBR322 ***(10,34)***, pTZ18U [a high-copy number plasmid vector ***(35,36)***], and "runaway" vectors whose replication rate increases at high growth temperature ***(37–39)***. These plasmids produce larger amounts of the enzyme than in an untransformed cell, making detection easier in a crude extract.

Several procedures have been devised for RecBCD enzyme purification ***(15,20,26,35,38,40–42)***. The purified enzyme is quite stable, with little loss of activity after at least 2 yr when stored at –80°C in 10 m*M* potassium phosphate, pH 6.9, 0.1 m*M* EDTA, 0.1 m*M* DTT, 50% (v:v) glycerol. The purified *E. coli* enzyme concentration can be determined from absorbance measurements, using $\varepsilon_{280} = 400{,}000\ M^{-1}\ \text{cm}^{-1}$ ***(26)***.

3.2.1. Preparation of Cell-Free Extract

The following procedure can be used to prepare a small cell-free extract for RecBCD assay ***(41)***.

1. Fill a 1.5-mL microcentrifuge tube with a culture of cells to be assayed for RecBCD enzyme activity (usually a midlog phase culture). Harvest the cells by sedimentation in a microcentrifuge (1 min, 8000 rpm).
2. Remove the culture medium supernatant by aspirating through a Pasteur pipet.
3. Resuspend the pellet in 100 µL of ice-cold 10 m*M* Tris-HCl, pH 7.5, 20% sucrose (w:v).
4. Freeze the suspension at –80 °C for at least 30 min, thaw, and put the tube on ice.
5. Add 2 µL of lysozyme (10 mg/mL in water) and 1 µL of 0.5 *M* β-mercaptoethanol. Put the tube on ice for 5 min.
6. Add 2.1 mL of 5 *M* NaCl. Put on ice, 25 min.
7. Put the tube in a 37°C water bath for 10 min to lyse the cells, and then chill on ice for 5 min.
8. Remove the cell debris by sedimentation in a microcentrifuge (5 min, maximum speed, at 4°C).

   Pipet off the supernatant (approx 90 µL or so), leaving the pellet behind (*see* **Note 15**).

9. Assay the lysate for ATP-dependent nuclease activity (see below) and total protein [by, e.g., the Bradford assay *(43)*], and analyze a sample by sodium dodecyl sulfate polyacrylamide gel electrophoresis (SDS-PAGE) (*see* **Note 16**).

3.3. Exonuclease Assay on Double-Stranded DNA

The assay measures the conversion of [^{3}H]-labeled dsDNA to small oligonucleotide fragments that are soluble in cold 10% trichloroacetic acid (TCA). The reaction conditions are:

1. 50 m*M* Tris-HCl, pH 8.5
2. 10 m*M* $MgCl_2$
3. 40 μ*M* ATP
4. 40 μ*M* (nucleotides) [^{3}H]DNA
5. 0.67 m*M* DTT
6. 0.2 mg/mL BSA

in a total volume of 20 μL (*see* **Note 17**).

1. Mix, in a 1.5-mL plastic microcentrifuge tube:
   a. 5 μL 4X reaction buffer
   b. 0.8 μL ATP (1.0 m*M* stock)
   c. [^{3}H]DNA (e.g., 1.6 μL of 500 μ*M* stock)
   water [20 μL – (vol enzyme to be added)] (*see* **Note 18**).
   Put the tube in a 37°C water bath for approx 5 min. Keep the stock solutions on ice.
2. Initiate the reaction by adding RecBCD enzyme, approx 1–5 μL of a crude extract, or 1 μL of purified enzyme diluted in the buffer described above. Mix and place at 37°C (*see* **Note 19**).
3. Quench the reaction after 10 min by adding 100 μL of ice-cold 10% TCA and 5 μL of calf thymus DNA (0.5 mg/mL). Mix and put the tube on ice for at least 5 min.
4. Spin 10 min in a microcentrifuge, maximum speed, at 4°C.
5. Remove 110 μL of the supernatant (do not disturb the (invisible) pellet of large [^{3}H]DNA at the bottom of the tube). Add to 90 μL of water and 3 mL of scintillation fluid, mix thoroughly, and count in the [^{3}H] channel of the scintillation counter.

The concentration of solubilized nucleotide residues in the reaction mixture (assuming that the sample volumes are as given in this procedure) can be calculated from:

$$[\text{acid soluble DNA}]\ (\mu M\ \text{nt}) = \frac{\text{cpm} - BG}{\text{cpm/nmol nt}} \times \frac{125\ \mu\text{L}}{110\ \mu\text{L}} \times \frac{1}{20\ \mu\text{L}} \times 10^3$$

where *BG* is the cpm in the reaction mixture or aliquot containing no enzyme (*see* **Note 20**).

4. Notes

1. RNase A is available from several sources, and may be heat-treated as in **ref.** ***(33)***, p. B.17, to inactivate any DNase that may be present.
2. BSA may precipitate from solutions stored at –20°C, and can be redissolved by mixing and gentle heating. BSA is not essential for these reactions, particularly with the purified RecBCD enzyme.
3. ATP is an acid with four ionizable groups ***(44)***. The ATP concentration in the stock should be determined from UV absorbance measurements [λ_{max} = 259 nm; ε_{259} = 15,400 M^{-1} cm^{-1} (**ref.** ***(33)***, p. B.11)]. The 0.1 *M* ATP is stable for months at –20 °C.
4. We have noted no deterioration in the effectiveness of a 100% TCA solution after several years use. Caution: TCA is a strong acid [pKa = 0.70 ***(45)***].
5. Calf thymus DNA is added to the quenched reaction mixtures to serve as a carrier and enhance the precipitation of large DNA fragments. The exact DNA concentration can be determined from absorbance measurements (A_{260} of 1 = 50 µg/mL, in a 1-cm pathlength cell), if desired.
6. Other nutrients may be required for a particular strain. For example, HB101 requires proline (0.1 g/L), and JM109 requires thiamine (0.1 g/L). A rich medium like LB broth should not be used.
7. Special care should be taken when working with milliCurie amounts of tritium. This includes:

   a. Clear a space on the lab bench. Move aside all equipment and supplies which will not be used for the [^{3}H]DNA prep.
   b. Place a new sheet of plastic-backed absorbent paper on the lab bench. Do all work over this paper and dispose of it when the procedure is completed.
   c. If possible, incubate the culture containing [^{3}H]thymidine in a confined incubator/shaker, rather than a common incubator room, in case of breakage.
   d. Dispose of the growth medium supernatant as liquid radioactive waste. A small amount of bleach may be added first. Do not autoclave the ^{3}H-containing medium.

8. Uridine inhibits thymidine phosphorylase, which would convert the [^{3}H]thymidine to thymine and prevent its incorporation into DNA ***(46)***.
9. The cell pellet can be stored frozen at –80°C if desired. Freezing and thawing aids the lysis but is not required.
10. The DNA will be difficult to pipet accurately if it is very large and/or concentrated. This makes it difficult to determine its concentration and specific radioactivity.
11. Dilute the DNA stock with TE8.0 so as to achieve A_{260} approx 0.5–1. Do duplicate measurements, and dilute the stock if it is too concentrated to measure accurately (*see* **Note 10**). The A_{260}/A_{280} = 1.95 for pure *E. coli* DNA ***(47)***.
12. Invert the vial several times to mix the aqueous sample thoroughly with the scintillation fluid and produce a clear homogeneous mixture. Count duplicate samples to be sure they agree well (*see* **Note 10**).

TCA quenches the scintillation counting slightly (approx 10%), when present at the same concentration as in a quenched reaction aliquot (**step 5, Subheading 3.3.**). For greatest accuracy, TCA should be included in the scintillation vial at the same concentration as in the quenched samples from reaction mixtures.

13. The procedure is for the plasmid pTZ19R [pUC plasmid replication origin ***(36)***]. pBR322 and related plasmids can also be amplified with chloramphenicol ***(48)***. This is not necessary for higher copy number plasmids such as pUC and derivatives.
14. The specific end structure of the linear DNA is not critical, as long as neither end has a single-stranded overhang longer than about 25 nt ***(20)***. Thus, plasmid DNA cleaved by any restriction endonuclease is a nuclease substrate for RecBCD. A typical restriction digest mixture would contain 100 µL of plasmid DNA, 12 µL of 10X restriction endonuclease reaction buffer, approx 10–20 U of restriction endonuclease, and sterile deionized water to 120 µL total volume. The progress of the digestion should be checked after 2–3 h at the appropriate temperature (usually 37°C) by running a small sample from the digest mixture on an agarose gel. The linearized DNA should be purified and isolated by phenol/chloroform extraction and ethanol precipitation and dissolved in 100 µL TE 8.0.
15. Sometimes the pellet is very loose and does not sediment well. Spin longer and/or pipet off less supernatant if this occurs.
16. Load 10–20 µg protein in each lane of a 7.5% polyacrylamide gel containing SDS (the extract usually contains about 0.5–1 mg/mL of total protein). The RecB and RecC proteins may be visible if a strain containing cloned genes is used. They are not visible in an extract of nonoverexpressing cells.
17. The best reaction conditions are those in which the enzyme produces mostly small DNA fragments: high Mg^{2+} (1–10 m*M*) and low ATP. The production of acid-soluble fragments is maximal at 40–60 µ*M* ATP, and is *lower* at higher ATP ***(15,40,49)***. The nuclease activity is readily detectable at pH 7–8.5. Under other conditions, and depending on the rate of DNA unwinding vs the frequency of nuclease cleavage, mostly large nuclease products may be formed. Fragments larger than approx 20 nt are acid-precipitable ***(32)*** and thus not detected in this assay. A nuclease reaction under these conditions would therefore give a falsely low signal.
18. Several important control mixtures should also be prepared:
    a. Leave out ATP. This measures the amount of DNA solubilized by nucleases other than RecBCD.
    b. Add 1 µL of dilution buffer alone, containing no enzyme, incubate for 10 min at 37°C, and quench with TCA. This sample gives the background of tritium label in the [^{3}H]DNA that is acid-soluble initially. This should be very low (<1% of the total cpm in the reaction). Subtract these background counts from those obtained from the reactions that contain enzyme.
    c. The total cpm in a reaction mixture or aliquot can be determined as follows: Add 100 µL of water and 5 µL of calf thymus DNA to a 20 µL reaction mixture prepared as in step 1, remove 110 µL (do not spin), and add to a scintillation vial containing 3 mL scintillation fluid and 90 µL of 10% TCA

(*see* **Note 12**). There will be at least 16,000 total cpm in a 20 μL reaction mixture by this assay procedure.

19. Several reactions can be done simultaneously by initiating them at 30- or 60-s intervals. If so, be sure to quench them at intervals as well (**step 3**), so that each reaction goes for only 10 min.

    The nuclease activity in an extract of wild-type cells is low, so 1–5 μL of undiluted extract should be added to the assay. The activity with a wild-type cell extract is about twofold greater with ATP than in its absence ***(15,34)***.

    The activity is somewhat greater in an extract of cells containing cloned recBCD genes. The nuclease activity in extracts of cells transformed with pFS11–04 ***(34)*** or pDJ05 [pTZ18U vector ***(35)***] is about 10–20-fold greater with ATP than without ***(34,50)***. Thus, the extract can be diluted approx fivefold for the assay. It is also a good practice to do assays with varied amounts of extract, to be sure that the observed activity is linear with amount of enzyme added. The maximum cpm solubilized in a reaction mixture should ideally be <50% of the total in the mixture.

    About 0.1 n*M* of purified RecBCD enzyme is sufficient for the ds-exonuclease assay. Nuclease activity is undetectable with the purified enzyme in the absence of ATP.

20. This procedure and the reaction volume given are for a fixed-time-point assay. To follow the time-course in a single reaction mixture, prepare enough reaction mixture for all aliquots (including a zero time-point, before the enzyme has been added), + 5–10 μL extra. Put 100 μL of 10% TCA into microcentrifuge tubes (one for each time point), and put them on ice. Remove a 20-μL aliquot from the reaction mixture, add it to the first tube containing TCA, add calf thymus DNA, mix, and put on ice. Initiate the reaction by adding RecBCD enzyme to the remaining reaction mixture, and start a timer. Remove 20 μL aliquots at the desired times and add to the remaining quench tubes. When all aliquots have been removed, centrifuge the quenched samples and count.

    Reaction time-courses with dsDNA are usually linear for only a short time (<5 min), although the extent of reaction is proportional to the enzyme concentration ***(40)***. The reaction with denatured DNA is more linear with time. The dsDNA can be heat-denatured by placing a tube in a boiling water bath for 5 min, and then transferring it to an ice-water bath. The reaction mixture for ssDNA can be similar to that used with dsDNA, except the ATP concentration should be greater [0.2 m*M* ATP; the apparent nuclease activity does not decline at high [ATP] in this assay ***(15)***]. The reaction with ssDNA requires a greater enzyme concentration than for the ds-exonuclease activity (approx fivefold greater) to obtain a good signal.

    RecBCD hydrolyzes about 2 ATP per basepair unwound and made acid soluble ***(25,51)***. ATP regeneration can be important in some cases, given that the total ATP concentration is rather low in a ds-exonuclease reaction. An example was the case of the RecBCD-K177Q mutant enzyme. ATP hydrolysis by this mutant is relatively fast compared to the nuclease, so, at low ATP (10 μ*M*), the ATP was used up and the nuclease reaction stopped prematurely ***(52)***.

An ATP regeneration system consists of 2 m*M* phosphoenolpyruvate (PEP) and pyruvate kinase (20 U/mL). PEP and pyruvate kinase (type II, approx 1800 U/mL, from rabbit muscle, as a suspension in 3.2 *M* ammonium sulfate) can be obtained from Sigma. Prepare a 100 m*M* PEP stock in 50 m*M* Tris-HCl, pH 7.5. Store frozen in aliquots at –20°C. The pyruvate kinase mixture should be shaken gently to resuspend and then approx 20 μL of the suspension transferred to a microcentrifuge tube. Spin for 5 min at 4°C, pipet off the supernatant, and redissolve the protein pellet in 54 μL of 50 m*M* Tris-HCl, pH 7.5, and 36 μL of PEP solution. Add 1 μL of this mixture to a 20 μL reaction.

References

1. Myers, R. S. and Stahl, F. W. (1994) κ and the RecBC D enzyme of *Escherichia coli*. *Annu. Rev. Genet.* **28,** 49–70.
2. Smith, G. R., Amundsen, S. K., Dabert, P., and Taylor, A. F. (1995) The initiation and control of homologous recombination in *Escherichia coli*. *Phil. Trans. R. Soc. Lond. B.* **347,** 13–20.
3. Kowalczykowski, S. C., Dixon, D. A., Eggleston, A. K., et al. (1994) Biochemistry of homologous recombination in *Escherichia coli*. *Microbiol. Rev.* **58,** 401–465.
4. Kooistra, J. and Venema, G. (1991) Cloning, sequencing, and expression of *Bacillus subtilis* genes involved in ATP-dependent nuclease synthesis. *J. Bacteriol.* **173,** 3644–3655.
5. El Karoui, M., Ehrlich, D., and Gruss, A. (1998) Identification of the lactococcal exonuclease/recombinase and its modulation by the putative Chi sequence. *Proc. Natl. Acad. Sci. USA* **95,** 626–631.
6. Kooistra, J., Haijema, B. J., and Venema, G. (1993) The *Bacillus subtilis addAB* genes are fully functional in *Escherichia coli*. *Mol. Microbiol.* **7,** 915–923.
7. Shemyakin, M. F., Grepachevsky, A. A., and Chestukhin, A. V. (1979) Properties of *Bacillus subtilis* ATP-dependent deoxyribonuclease. *Eur. J. Biochem.* **98,** 417–423.
8. Smith, G. R. (1990) RecBCD Enzyme, in *Nucleic Acids and Molecular Biology,* vol. 4 (Eckstein, F. and Lilley, D. M. J., eds.), Springer-Verlag, Berlin, Heidelberg, pp. 78–98.
9. Muskavitch, K. M. T. and Linn, S. (1981) *recBC*-like enzymes: Exonuclease V deoxyribonucleases, in *The Enzymes,* vol. 14 (Boyer, P.D., ed.), Academic Press, New York, pp. 233–250.
10. Ponticelli, A. S., Schultz, D. W., Taylor, A. F., and Smith, G. R. (1985) Chi-dependent DNA strand cleavage by RecBC enzyme. *Cell* **41,** 145–151.
11. Dixon, D. A. and Kowalczykowski, S. C. (1993) The recombination hotspot χ is a regulatory sequence that acts by attenuating the nuclease activity of the *E. coli* RecBCD enzyme. *Cell* **73,** 87–96.
12. Oishi, M. (1969) An ATP-dependent deoxyribonuclease from *Escherichia coli* with a possible role in genetic recombination. *Proc. Natl. Acad. Sci. USA* **64,** 1292–1299.
13. Barbour, S. D. and Clark, A. J. (1970) Biochemical and genetic studies of recombination proficiency in *Escherichia coli*, I. Enzymatic activity associated with *recB*+ and *recC*+ genes. *Proc. Natl. Acad. Sci. USA* **65,** 955–961.

14. Wright, M., Buttin, G., and Hurwitz, J. (1971) The isolation and characterization from *Escherichia coli* of an adenosine triphosphate-dependent deoxyribonuclease directed by *rec B, C* genes. *J. Biol. Chem.* **246,** 6543–6555.
15. Goldmark, P. J. and Linn, S. (1972) Purification and properties of the *recBC* DNase of *Escherichia coli* K–12. *J. Biol. Chem.* **247,** 1849–1860.
16. Chen, H.-W., Randle, D. E., Gabbidon, M., and Julin, D. A. (1998) Functions of the ATP hydrolysis subunits (RecB and RecD) in the nuclease reactions catalyzed by the RecBCD enzyme from *Escherichia coli. J. Mol. Biol.* **278,** 89–104.
17. Rosamond, J., Telander, K. M., and Linn, S. (1979) Modulation of the action of the *recBC* enzyme of *Escherichia coli* K–12 by Ca^{2+}. *J. Biol. Chem.* **254,** 8646–8652.
18. MacKay, V. and Linn, S. (1974) The mechanism of degradation of duplex deoxyribonucleic acid by the *recBC* enzyme of *Escherichia coli* K–12. *J. Biol. Chem.* **249,** 4286–4294.
19. Taylor, A. and Smith, G. R. (1980) Unwinding and rewinding of DNA by the RecBC enzyme. *Cell* **22,** 447–457.
20. Taylor, A. F. and Smith, G. R. (1985) Substrate specificity of the DNA unwinding activity of the RecBC enzyme of *Escherichia coli. J. Mol. Biol.* **185,** 431–443.
21. Lohman, T. M. and Bjornson, K. P. (1996) Mechanism of helicase-catalyzed DNA unwinding. *Annu. Rev. Biochem.* **65,** 169–214.
22. Eggleston, A. K. and Kowalczykowski, S. C. (1993) Biochemical characterization of a mutant recBCD enzyme, the $recB^{2109}CD$ enzyme, which lacks κ-specific, but not non-specific, nuclease activity. *J. Mol. Biol.* **231,** 605–620.
23. Muskavitch, K. M. T. and Linn, S. (1982) A unified mechanism for the nuclease and unwinding activities of the *recBC* enzyme of *Escherichia coli. J. Biol. Chem.* **257,** 2641–2648.
24. MacKay, V. and Linn, S. (1976) Selective inhibition of the DNase activity of the *recBC* enzyme by the DNA binding protein from *Escherichia coli. J. Biol. Chem.* **251,** 3716–3719.
25. Korangy, F. and Julin, D. A. (1994) Efficiency of ATP Hydrolysis and DNA unwinding by the RecBC enzyme from *Escherichia coli. Biochemistry* **33,** 9552–9560.
26. Roman, L. J. and Kowalczykowski, S. C. (1989) Characterization of the helicase activity of the *Escherichia coli* RecBCD enzyme using a novel helicase assay. *Biochemistry* **28,** 2863–2873.
27. Roman, L. J., Eggleston, A. K., and Kowalczykowski, S. C. (1992) Processivity of the DNA helicase activity of *Escherichia coli* recBCD enzyme. *J. Biol. Chem.* **267,** 4207–4214.
28. Weiss, B. (1981) Exodeoxyribonucleases of *Escherichia coli*, in *The Enzymes,* vol. XIV (Boyer, P. D., ed.), Academic Press, New York, pp. 203–231.
29. Anderson, D. G. and Kowalczykowski, S. C. (1997) The recombination hot spot κ is a regulatory element that switches the polarity of DNA degradation by the RecBCD enzyme. *Genes Dev.* **11,** 571–581.
30. McKittrick, N. H. and Smith, G. R. (1989) Activation of Chi recombinational hotspots by RecBCD-like enzymes from enteric bacteria. *J. Mol. Biol.* **210,** 485–495.

31. Sourice, S., Biaudet, V., El Karoui, M., et al. (1998) Identification of the Chi site of *Haemophilus influenzae* as several sequences related to the *Escherichia coli* Chi site. *Mol. Microbiol.* **27,** 1021–1029.
32. Cleaver, J. E. and Boyer, H. W. (1972) Solubility and dialysis limits of DNA oligonucleotides. *Biochim. Biophys. Acta* **262,** 116–124.
33. Sambrook, J., Fritsch, E. F. and Maniatis, T. (1989) *Molecular Cloning: A Laboratory Manual*, Second Edition, Cold Spring Harbor Laboratory Press, Cold Spring Harbor, NY.
34. Sasaki, M., Fujiyoshi, T., Shimada, K., and Takagi, Y. (1982) Fine structure of the *recB* and *recC* gene region of *Escherichia coli. Biochem. Biophys. Res. Commun.* **109,** 414–422.
35. Korangy, F. and Julin, D. A. (1992) Alteration by site-directed mutagenesis of the conserved lysine residue in the ATP-binding consensus sequence of the RecD subunit of the *Escherichia coli* RecBCD enzyme. *J. Biol. Chem.* **267,** 1727–1732.
36. Mead, D. A., Szczesna-Skorupa, E., and Kemper, B. (1986) Single-strand DNA "blue" T7 promoter plasmids: a versatile tandem promoter system for cloning and protein engineering. *Protein Eng.* **1,** 67–74.
37. Amundsen, S. K., Taylor, A. F., Chaudhury, A. M., and Smith, G. R. (1986) *recD*: the gene for an essential third subunit of exonuclease V. *Proc. Natl. Acad. Sci. USA* **83,** 5558–5562.
38. Julin, D. A. and Lehman, I. R. (1987) Photoaffinity labelling of the recBCD enzyme of *Escherichia coli* with 8-azidoadenosine 5'- triphosphate. *J. Biol. Chem.* **262,** 9044–9051.
39. Finch, P. W., Storey, A., Brown, K., et al. (1986) Complete nucleotide sequence of *recD,* the structural gene for the α subunit of Exonuclease V of *Escherichia coli. Nucleic Acids Res.* **14,** 8583–8594.
40. Eichler, D. C. and Lehman, I. R. (1977) On the role of ATP in phosphodiester bond hydrolysis catalyzed by the *rec*BC deoxyribonuclease of *Escherichia coli. J. Biol. Chem.* **252,** 499–503.
41. Dykstra, C. C., Palas, K. M., and Kushner, S. R. (1984) Purification and characterization of Exonuclease V from *Escherichia coli* K-12. *Cold Spring Harbor Symp. Quant. Biol.* **XLIX,** 463–467.
42. Masterson, C., Boehmer, P. E., McDonald, F., et al. (1992) Reconstitution of the activities of the RecBCD holoenzyme of *Escherichia coli* from the purified subunits. *J. Biol. Chem.* **267,** 13,564–13,572.
43. Bradford, M. M. (1976) A rapid and sensitive method for the quantitation of microgram quantities of protein utilizing the principle of protein-dye binding. *Anal. Biochem.* **72,** 248–254.
44. O'Sullivan, W. J. and Smithers, G. W. (1979) Stability constants for biologically important metal-ligand complexes. *Methods Enzymol.* **63,** 294–336.
45. Weast, R. C., ed. (1987) *CRC Handbook of Chemistry and Physics*. CRC, Boca Raton, FL, p. D–162.

46. Kornberg, A. and Baker, T. A. (1991) *DNA Replication*, 2nd Ed., W. H. Freeman, New York, p. 89.
47. Rodriguez, R. L. and Tait, R. C. (1983) *Recombinant DNA Techniques: An Introduction,* Benjamin/Cummings, Menlo Park, CA, p. 43.
48. Halford, S. E. and Goodall, A. J. (1988) Modes of DNA cleavage by the *Eco*RV restriction endonuclease. *Biochemistry* **27,** 1771–1777.
49. Karu, A. E., MacKay, V., Goldmark, P. J., and Linn, S. (1973) The *recBC* deoxyribonuclease of *Escherichia coli* K–12. *J. Biol. Chem.* **248,** 4874–4884.
50. Hsieh, S. and Julin, D. A. (1992) Alteration by site-directed mutagenesis of the conserved lysine residue in the consensus ATP-binding sequence of the RecB protein of *Escherichia coli. Nucleic Acids Res.* **20,** 5647–5653.
51. Roman, L. J. and Kowalczykowski, S. C. (1989) Characterization of the adenosinetriphosphatase activity of the *Escherichia coli* RecBCD enzyme: relationship of ATP hydrolysis to the unwinding of duplex DNA. *Biochemistry* **28,** 2873–2881.
52. Korangy, F. and Julin, D. A. (1992) Enzymatic effects of a lysine-to-glutamine mutation in the ATP-binding consensus sequence in the RecD subunit of the RecBCD enzyme from *Escherichia coli. J. Biol. Chem.* **267,** 1733–1740.

9

Holliday Junction Branch Migration and Resolution

RuvA, RuvB, and RuvC from the Hyperthermophile Thermotoga maritima

Susana Gonzalez and James G. Wetmur

1. Introduction

Two types of general methods are detailed in this chapter. The first involves formation of a Holliday junction substrate that can be used to measure both helicase-dependent Holliday junction branch migration and Holliday junction resolution driven by Ruv proteins from any bacterial species, whether mesophilic or thermophilic. The second involves the purification of Ruv proteins from the thermophile, *Thermotoga maritima,* by procedures applicable to the purification of thermostable proteins from a variety of thermophilic genera. The specific assays detail the measurement of branch migration and resolution of the Holliday junction substrate by *T. maritima* Ruv proteins.

The Holliday junction is a key intermediate in homologous recombination. The X-shaped junction is formed at the intersection of two homologous duplex DNAs with one strand of each DNA duplex crossing over to the second duplex. In prokaryotes such as *Escherichia coli*, Holliday junctions may be formed by the actions of RecA, RecBCD, and SSB *(1)*. Subsequently, RuvB, together with RuvA, forms a complex on the Holliday junction that catalyzes branch migration *(2)*. Although RecA itself can catalyze branch migration through mismatched sequences, both RuvA and RuvB are needed specifically for bypass of insertions >100 bp and, in general, to speed up branch migration *(3)*. The tetrameric RuvA protein binds specifically to the "square" planar form of the Holliday junction *(4,5)* and can form a complex with RuvB in solution. RuvB is a DNA-dependent ATPase *(6)*, which forms hexameric rings on duplex DNA

From: *Methods in Molecular Biology, vol. 152: DNA Repair Protocols: Prokaryotic Systems*
Edited by: P. Vaughan © Humana Press Inc., Totowa, NJ

(7). High concentrations of RuvB will promote branch migration, bound RecA displacement, and closed circular DNA unwinding in vitro in the presence of ATP and high concentrations of Mg^{2+}, indicating that RuvB is the translocating motor driving Holliday junction branch migration. At low concentrations of Mg^{2+}, RuvA is required to direct RuvB to and maintain RuvB on the Holliday junction substrate during branch migration ***(8)***.

Following branch migration catalyzed by RuvAB, Holliday junction resolution is accomplished by RuvC. The dimeric structure of the RuvC protein ***(9)*** is consistent with its ability to cleave across the Holliday junction by dual incision at base-specific contacts, a mechanism similar to that observed with certain restriction endonucleases. The recognition and cleavage site for *E. coli* RuvC is (A/T)TT $\downarrow$ (G/C) ***(10)***. RuvC works with RuvAB in the late steps of homologous recombination ***(11–13)***. RuvC forms a complex with RuvAB at a Holliday junction, and enhanced RuvC cleavage depends on RuvAB-mediated ATP hydrolysis.

Biochemical studies employing thermostable proteins overexpressed in *E. coli* have unique advantages. Purification of thermostable proteins to homogeneity is greatly simplified by the use of a heating step to denature and subsequently precipitate host proteins. The wide range of available assay temperatures permits studies at very low turnover rates. Trace host protein contaminants are unlikely to have any enzymatic activities at the elevated temperatures used for assays. Because thermostability is rarely affected significantly by site directed mutagenesis, mutant proteins may be purified using the same protocols. Thus, thermostable proteins also are good candidates for structural studies.

The genomic sequences for *T. maritima* RecA and RuvB were cloned using degenerate and inverse PCR ***(14,15)***. Recently, the complete genomic sequences of several thermophiles have been reported, including *T. maritima* ***(16)***, greatly simplifying the cloning of RuvA and RuvC (Gonzalez, S., Rosenfeld, A., Szeto, D., et al., manuscript in preparation). The three *T. maritima* Ruv proteins have been expressed in *E. coli* and purified to homogeneity. The *T. maritima* RuvAB catalyzes Holliday junction branch migration at 70°C leading to products containing regions of nonhomology. Either ATP or ATPγS hydrolysis will serve as the energy source. *T. maritima* RuvC resolves Holliday junctions at 70°C. Remarkably, the cleavage site [(A/T)TT $\downarrow$ (G/C)] is identical to the preferred cleavage site for *E. coli* RuvC. Phylogenetic analysis based on RecA and RuvB has shown that thermophilic bacteria such as *T. maritima,* representative mesophilic Gram negative bacteria, and representative mesophilic Gram-positive bacteria are equally divergent from one another. The conservation of function strengthens the argument for using the *T. maritima* proteins for biochemical and structural studies of the RuvAB helicase and RuvC resolvase activities.

2. Materials

2.1. Reagents and Procedures

1. All synthetic oligodeoxynucleotide primers for PCR were ordered from IDT, Inc. (www.idtdna.com). The primers were synthesized on automated instruments.
2. All enzymes were purchased from New England Biolabs (www.neb.com), except where noted and used as indicated by the manufacturer.
3. All DNA manipulations used standard techniques and procedures *(17)*.
4. LB medium: 10 g/L bacto tryptone (DIFCO), 5 g/L bacto yeast extract (DIFCO), 10 g/L NaCl, Deionized H_2O to 1 L. Adjust pH to 7.0 with 1 *M* NaOH.
5. LB medium supplemented with 50 µg/mL ampicillin and 25 µg/mL chloramphenicol.
6. Cell resuspension buffer: 50 m*M* Tris-HCl, pH 8.0, 10 m*M* ethylenediaminetetraacetic acid (EDTA), 1 m*M* dithiothreitol (DTT) and 1m*M* phenyl methyl sulfonyl flouride (PMSF).
7. 1 *M* $(NH_4)_2SO_4$.
8. 20 m*M* Tris-HCl, pH 9.0.
9. TE: 10 m*M* Tris-HCl, 1 m*M* EDTA, pH 7.8 / pH 8.0.
10. 0.1 *M* NaOAc in TE, pH 7.8.
11. 2 *M* NaCl and 0.2 *M* NaCl in TE, pH 8.0.
12. 0.1 *M* KCl and 0.2 *M* KCl in TE, pH 8.0.
13. 2 *M* guanidine HCl in TE, pH 7.8.
14. TBE buffer: 89 m*M* Tris-HCl, 89 m*M* borate, 1 m*M* EDTA, pH 8.
15. TAE buffer: 0.04 *M* Tris-acetate, 1 m*M* EDTA, pH 8.0.
16. HiTrap Q anion-exchange column and HiTrap Blue column (Amersham-Pharmacia, www.apbiotech.com).

2.2. Holliday Junction Substrates

2.2.1. Principles of Reassociation-Mediated Holliday Junction Formation

The Holliday junction substrate used in this work was formed by denaturation and reassociation of two duplex reactants, each consisting mostly of an inverted repeat. Reassociation-mediated Holliday junction formation is depicted in **Fig. 1**. Reactants were formed by head to head ligation of PCR products differing only in the sequences of the PCR primers (**Fig. 1**, Panel **A**). Each head primer contained a different restriction endonuclease recognition sequence capable of producing compatible ends for ligation. Each tail primer contained a unique 5' sequence of approximately 30 nt, depicted as an open box. Denaturation and cooling to a temperature just below the melting temperature converts the inverted repeat reactant strands into hairpins (**Fig. 1**, Panel **B**). Annealing the unique tail sequences leads to a transient Holliday junction structure (**Fig. 1**, Panel **B**). Spontaneous branch migration leads to reformation of the reactant (**Fig. 1**, Panel **A**). The use of an equimolar concentration of two reactants formed using templates differing at one site will lead to a 50% yield

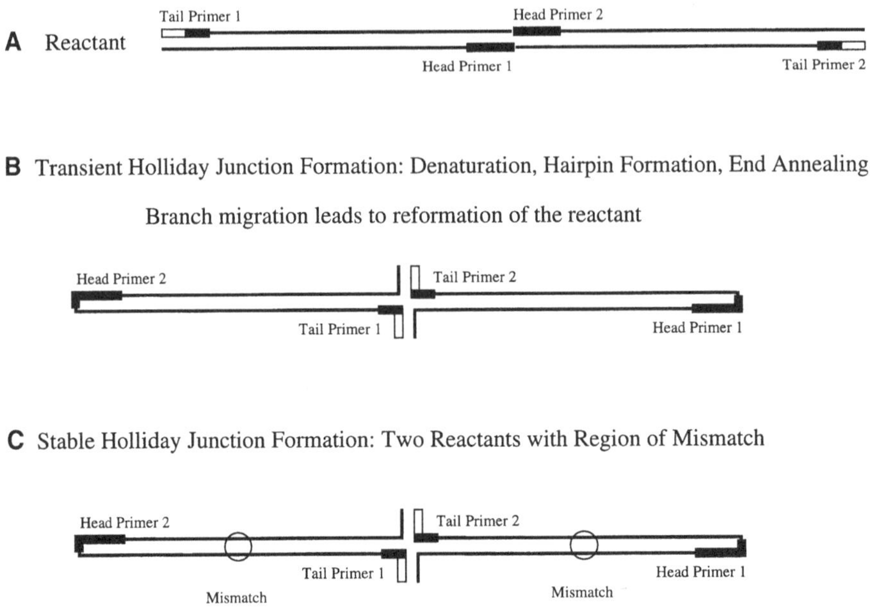

Fig. 1. Formation of stable Holliday junctions. Panel **A**: Structure of an inverted-repeat reactant formed by head to head ligation of PCR products derived from the same template. The white blocks on the tail primers denote unique sequences. Panel **B**: Hairpins formed by denaturation and cooling a reactant. The two hairpin strands are depicted with the complementary tail sequences aligned for initiation of a Holliday junction. Panel **C**: Hairpins formed by denaturation and cooling two reactants derived from two templates differing at one locus. The mismatch region occurs when the two strands are derived from different reactants.

of stable Holliday structures because branch migration is impeded by the region of mismatch (**Fig. 1**, Panel **C**).

2.2.2. PCR Products for Reassociation-Mediated Holliday Junction Formation

In the example below, pUC19 derivatives were used as PCR templates to form the repeat units of the reactants (**Fig. 1**, Panel **A**). In pUC19GC, the polylinker of pUC19 was modified at the *Bam*HI site, substituting a C for the T to create the sequence GGA**C**CC. pUC19Δ3 was formed by introducing a T and two Cs into the pUC19GC polylinker sequence to yield GGA**TCC**CCC and reconstitute the *Bam*HI site. Template concentrations for PCR were 5 μg/mL.

Both PCR primers for creating a 382-bp repeat unit included one of the two *Pvu*II sites in pUC19. The primer 5'-ggccagatctggccccatggcgaaggcctc TACG-CCAGCTTGG-CGAAAGGG-3' (Tail Primer 1) was used with 5'-GCG-

CCAC**CTGA**GTGGAATGC-AGCCTGGCACGACAGG-3' (Head Primer 1) to form one repeat unit. The primers for the second repeat unit were 5'-gcgagctagcggccgatatcgcggccgcaaTACGCCAGCTGGC-GAAAGGG-3' (Tail Primer 2) and 5'-GCGCCAC**TCAG**GTGGAATGCAGCTGGCACGACAGG-3' (Head Primer 2). The unique sequences in the Tail Primers are indicated in lower case. The *Bst*XI sites in the Head Primers are underlined. The unique and selfcomplementary sticky end sequences are shown in boldface. PCR amplifications were carried out using standard reagent concentrations *(18)* using cycles of 30 s, 30 s and 60 s at 94°C, 55°C, and 72°C, respectively.

PCR products were purified with a Qiagen PCR purification kit (www.qiagen.com) according to the manufacturer's instructions. The purified PCR products were digested overnight at 55°C with *Bst*XI in NEB3 Buffer (50 m*M* Tris-HCl, pH 7.9, 10 m*M* $MgCl_2$, 100 m*M* NaCl, 1 m*M* DTT, New England Biolabs), heated to 65°C for 20 min to inactivate *Bst*XI, and washed on Ultrafree-MC filters (Millipore, www.millipore.com) with TE, pH 7.8 to remove the short fragments. To form an inverted repeat reactant (**Fig. 1**, Panel **A**), an equimolar mixture of the *Bst*XI-digested PCR products, produced using the two primer sets on the same template, was ligated overnight with T4 DNA ligase in buffer containing ATP. The ligated products were separated from unligated products by electrophoresis on a 2% agarose gel in TAE buffer, followed by excision from the gel and extraction from the agarose using an agarose gel DNA extraction kit (Roche Molecular Biochemicals, biochem. roche.com). The agarose gel and the running buffer contained 0.5 µg/mL ethidium bromide (*see* **Note 1**). The reactants were further purified by phenol:chloroform extraction and ethanol precipitation, resuspended in TE and stored at 4°C until further use.

2.2.3. Other Holliday Junction Substrates

Ruv protein substrates can be formed by using equimolar mixtures of partially complementary synthetic oligonucleotides to form Holliday junctions *(19)*. These substrates were not used for studies with the thermostable Ruv proteins because of their instability at the high temperatures required for optimal activity of the proteins.

Experimental systems have been developed to study the interactions of *E. coli* Ruv proteins and RecA. These studies employed the recombination intermediates that were formed during the DNA strand exchange reaction between gapped circular DNA and ^{32}P-3'-end-labeled homologous linear-duplex DNA upon addition of RecA *(11,20)*. These recombination intermediates (α structures) contained Holliday junctions that were substrates for the Ruv proteins. This type of substrate was not used with the thermostable Ruv proteins because of the inefficiency of substrate formation by the homologous thermostable *T. maritima* RecA protein.

2.3. Bacterial Strains Expressing Recombinant T. maritima Ruv Proteins

E. coli strain BL21(DE3) pLysS, *endA*⁻ [BL21-Gold (DE3) pLysS] (Stratagene, www.stratagene.com) were used to carry the plasmid pET11c (Novagen). BL21(DE3) is a derivative of *E. coli* B carrying an IPTG-inducible T7 RNA polymerase gene of the prophage DE3. The plasmid pET11c contains a T7 RNA polymerase promoter positioned to permit expression from an inserted gene under the control of T7 RNA polymerase. The pLysS plasmid encodes lysozyme, which facilitates release of expressed recombinant proteins by freeze-thaw lysis of the cells. The *endA*⁻ mutation inactivates the periplasmic space enzyme endonuclease I that cuts double-stranded DNA nonspecifically. *E. coli* endonuclease I is thermostable and, to some degree, thermoactive. The cloning of *T. maritima ruvA*, *ruvB,* and *ruvC* into pET11c was accomplished by amplification of the genes with appropriate PCR primers containing introduced *Nde*I and either *Bam*HI or *Bgl*II restriction endonuclease sites. The PCR products were digested and ligated into the *Nde*I and *Bam*HI sites of pET11c. After ligation, the vectors were transformed into BL21-Gold (DE3) pLysS. The sequences of the plasmid inserts were verified.

3. Methods

3.1. Substrate Formation

Reactants (Fig. 1, Panel A) were prepared as described in **Subheading 2.2.2.** The PCR template for preparation of one reactant was pUC19GC. The second reactant was based on amplification products of pUC19Δ3 using the same primer sets. In typical reactions, stable Holliday junctions were formed from equal quantities of each reactant in NEB3 Buffer. The mixtures were incubated to 100°C for 1 min to separate the strands, annealed at 85°C for a maximum of 2 min to allow hairpin formation (**Fig. 1**, Panels **B**), and further annealed for a maximum of 60 min at 70°C to permit formation of stable Holliday junctions at a 50% yield (**Fig. 1**, Panel **C**). RuvA, RuvB, or both are added in at least three-fold stoichiometric excess to the stable Holliday junctions for analysis of ATP-dependent branch migration as described below in **Subheading 3.3.** There are 12 RuvB monomers (2 hexamers) per junction and 4 or 8 RuvA monomers (1 or 2 hexamers) per junction. The reactant concentrations in typical reactions were 30 n*M* each in a final volume of 30 µL.

3.2. T. maritima Ruv Protein Expression and Purification

3.2.1. T. maritima RuvA, RuvB, and RuvC Expression and Partial Purification

1. Grow the bacteria containing the pET11c expression plasmid for RuvA, RuvB, or RuvC in LB medium supplemented with 50 µg/mL ampicillin and 25 µg/mL chloramphenicol to A_{600} >0.5 (*see* **Note 2**).

2. Induce the cultures with 1m*M* IPTG.
3. Grow for an additional 4–6 h.
4. Collect the cells by centrifugation. The cells may be stored frozen overnight at –80°C.
5. Resuspend the cells in 1 mL 50 m*M* Tris-HCl, pH 8.0, 10 m*M* EDTA, 1 m*M* DTT, and 1 m*M* PMSF.
6. Freeze and thaw three times.
7. Sonicate the samples as necessary to reduce the viscosity caused by the bacterial DNA (*see* **Note 3**).
8. Clarify by microcentrifugation (*see* **Note 4**).
9. Transfer the supernatant to a new tube and add 1 *M* $(NH_4)_2SO_4$ to a final concentration of 0.3 *M* (*see* **Note 5**).
10. Heat the samples at 75°C for 15 min to denature *E. coli* proteins.
11. Place on ice for 30 min to aggregate *E. coli* proteins. The solution will look milky.
12. Clarify by microcentrifugation at 4°C for 15 min.
13. Transfer the supernatant to a new tube or tubes.
14. Crude RuvA, RuvB, or RuvC may be stored at 4°C for up to 2 d prior to purification.

3.2.2. T. maritima *RuvA, RuvB, and RuvC Purification to Homogeneity*

Protein purification may be accomplished by two chromatographic steps. A 5-mL HiTrap Q anion-exchange column (Amersham-Pharmacia, www.apbiotech.com) was used first because of its high capacity. A 5-mL HiTrap Blue column was then employed because it frequently binds proteins involved in nucleic acid metabolism very tightly and because its negative charge permits removal of any residual nucleic acids in the wash solutions. The manufacturer's protocols should be followed for proper equilibration of the columns prior to loading the proteins.

3.2.2.1. *T. maritima* RuvA

1. Load the crude RuvA directly onto a HiTrap Q column.
2. Wash with 100 mL 20 m*M* Tris-HCl, pH 9.0.
3. Elute in 0.2 *M* NaCl in 20 m*M* Tris-HCl, pH 9.0.
4. Identify RuvA fractions by sodium dodecyl sulfate-polyacrimide gel electrophoresis SDS-PAGE (*see* **Note 6**).
5. Dilute appropriate fractions five-fold into 0.1 *M* NaOAc in TE, pH 7.8
6. Load onto a HiTrap Blue column.
7. Wash with 100 mL 0.1 *M* NaOAc in TE, pH 7.8.
8. Elute with 2 *M* NaCl in TE, pH 8.0.
9. Identify RuvA fractions by SDS-PAGE.
10. Dialyze overnight at room temperature into 0.2 *M* KCl in TE, pH 8.0

3.2.2.2. *T. maritima* RuvB

1. Dilute crude RuvB 10-fold into 20 m*M* Tris-HCl, pH 9.0
2. Load onto a HiTrap Q column.

3. Wash with 100 mL 20 m*M* Tris-HCl, pH 9.0.
4. Elute in 0.2 *M* NaCl in 20 m*M* Tris-HCl, pH 9.0.
5. Identify RuvB fractions by SDS-PAGE (*see* **Note 6**).
6. Dilute appropriate fractions fivefold into 0.1 *M* NaOAc in TE, pH 7.8
7. Load onto a HiTrap Blue column.
8. Wash with 100 mL 0.1 *M* NaOAc in TE, pH 7.8.
9. Elute with 2 *M* guanidine HCl in TE, pH 7.8.
10. Identify RuvB fractions by SDS-PAGE.
11. Dialyze overnight at room temperature into 0.2 *M* KCl in TE, pH 8.0

3.2.2.3. *T. maritima* RuvC

1. Dilute crude RuvC 10-fold into 20 m*M* Tris-HCl, pH 9.0
2. Load onto a HiTrap Q column.
3. Wash with 100 mL 20 m*M* Tris-HCl, pH 9.0.
4. Elute in 0.2 *M* NaCl in 20 m*M* Tris-HCl, pH 9.0.
5. Identify RuvC fractions by SDS-PAGE (*see* **Note 6**).
6. Dilute appropriate fractions five-fold into 0.1 *M* NaOAc in TE, pH 7.8.
7. Load onto a HiTrap Blue column.
8. Wash with 100 mL 0.1 *M* NaOAc in TE, pH 7.8.
9. Wash with 50 mL 2.0 *M* NaCl in TE, pH 7.8.
10. Elute with 2 *M* guanidine HCl in TE, pH 7.8.
11. Identify RuvC fractions by SDS-PAGE.
12. Dialyze overnight at room temperature into 0.1 *M* KCl in TE, pH 8.0.

Typical yields of *T. maritima* Ruv proteins are 0.2 mg/10^{11} cells, or approx 2.5% of the initial protein content of the cells. Protein purity may be examined by SDS-PAGE. Absorbance spectra should be taken to confirm a 280 nm maximum, indicating efficient removal of nucleic acids. The Ruv proteins may be concentrated as needed with either centricon-10 or centricon-30 filters (Amicon, www.millipore.com).

3.3. Analysis of Branch Migration

1. Form stable Holliday junctions in 6 tubes as described above in **Subheading 3.1.** Each volume should be less than 30 μL so that after the additions described as follows, the final concentration of stable Holliday junctions (50% yield) will be 30 n*M* in the final volume of 30 μL in NEB3 buffer.
2. Additions to tubes:
   a. Tube 1: RuvA to 1 μ*M* monomer plus 0.2 μ*M* ATP.
   b. Tube 2: RuvB to 1 μ*M* monomer plus 0.2 μ*M* ATP.
   c. Tube 3: RuvA and RuvB to 1 μ*M* monomers (No ATP).
   d. Tube 4: RuvA and RuvB to 1 μ*M* monomers plus 0.2 μ*M* ATP.
   e. Tube 5: RuvA and RuvB to 1 μ*M* monomers plus 0.2 μ*M* ATPγS.
   f. Tube 6: 0.2 μ*M* ATP (No RuvA or RuvB) as control.
3. Incubate at 70°C for 60 min.

4. *Optional*: Cut the products with one or more of the many restriction endonucleases that function well in NEB3 buffer (*see* **Notes 7** and **8**).
5. Add SDS to 0.1%.
6. Incubate at 50°C for 10 min. This procedure inactivates all the enzymes and disrupts all DNA-protein interactions.
7. Separate the products on a native 6% polyacrylamide gel in TBE buffer.
8. Stain the gel with ethidium bromide and examine by fluorography.

The denaturation and annealing of the two reactants leads to transient Holliday junctions formed from fully complementary strands (50% yield) and to stable Holliday junctions formed from complementary but mismatched strands (50% yield). Spontaneous branch migration of the transient Holliday junctions leads to reformation of the reactants. The stable Holliday junctions do not produce a useful band on the polyacrylamide gel. Thus, in the absence of RuvAB-directed branch migration, the only visible band should be the reactant band. Tubes 1–3 lack either RuvA, RuvB, or its energy source; control tube 6 lacks both RuvA and RuvB. Energy-dependent branch migration leads to formation of a heteroduplex reactant with two mismatched regions. For the substrate described above, the mismatched regions contain trinucleotide bulge loops that decrease the mobility relative to that of the homoduplex reactants. In the presence of RuvA, RuvB and an energy source (Tubes 4–5), homoduplex and heteroduplex reactant bands of similar intensity should be observed. Both ATP and ATPγS are hydrolyzed by *T. maritima* RuvB ***(15)***.

This branch migration assay may be used to determine the time course of energy-dependent branch migration. The function of native and mutant RuvA and RuvB proteins from any species may be examined as a function of temperature and solvent. Clearly the substrate described here may be modified to examine a wide variety of mismatch sequences.

3.4. Analysis of Resolution

1. Label one reactant (e.g., pUC19GC inverted repeat) using polynucleotide kinase [γ- ^{32}P]ATP.
2. Heat to 65°C for 20 min to inactivate the kinase.
3. Wash on Ultrafree-MC filters with magnesium-free NEB3 to exchange the buffer and remove the excess [γ- ^{32}P]ATP.
4. Form stable Holliday junctions in three tubes as described above in **Subheading 3.1.** Each volume should be less than 30 µL so that after the additions below the final concentration of stable Holliday junctions (50% yield) will be 30 n*M* in the final volume of 30 µL in NEB3 buffer.
5. Additions to tubes:
   a. Tube 1: RuvC to 1 µ*M* monomer plus $MgCl_2$ to 1 µ*M*.
   b. Tube 2: RuvC to 1 µ*M* monomer.
   c. Tube 3: $MgCl_2$ to 1 µ*M*.

6. Incubate at 70°C for up to 3 h.
7. Add SDS to 0.1%.
8. Incubate at 50°C for 10 min.
9. Three labeled marker DNAs may be prepared by digestion of the same labeled reactant with *Eco*RI, *Ava*II, and *Hind*III. Labeled sequencing ladders may be produced by PCR cycle sequencing using the same template (e.g., pUC19GC) and one labeled primer.
10. Separate the products on a DNA sequencing gel.
11. Examine by autoradiography to determine the location and extent of cleavage.

4. Notes

1. Caution: Because ethidium bromide may be activated to form a carcinogen, gloves should be worn when handling all gels and solutions containing ethidium bromide. Care must be taken to minimize UV damage when excising the band from the gel.
2. The ampicillin maintains the expression plasmid and the chloramphenicol maintains the compatible pLysS plasmid. The cultures may be grown overnight, diluted 1:100 the following morning into 250 mL of the same medium and grown until reaching an $A_{600} > 0.5$.
3. Before sonication, the high viscosity may be seen as slow liquid flow following inversion of a sample tube. A typical sonication procedure for a 1-mL sample on ice is three 10 s pulses spaced by a minimum of 30 s to avoid overheating the sample. After sonication and microcentrifugation, the liquid flow should be indistinguishable from buffer.
4. Up to and including this step, this workup would apply to most expressed proteins, whether from mesophilic or thermophilic organisms.
5. The high salt will prevent nonspecific binding of the thermostable proteins with other proteins or nucleic acids.
6. The sizes of the proteins following SDS-PAGE are RuvA: 20 kDa, RuvB: 37 kDa and RuvC: 18 kDa.
7. More specific structural data may be obtained by restriction endonuclease digestion of the stable Holliday junctions and branch migration products. For example, *Ava*II will cut pUC19GC and not pUC19Δ3, whereas *Bam*HI will cut pUC19Δ3 and not pUC19GC at the mismatch position. Heteroduplex reactants are refractory to cleavage.
8. For the reactants described above, simultaneous cleavage by *Ava*II and *Bam*HI cuts the approx 750 bp homoduplex reactants into a long inverted repeat containing the *Bst*XI site of approx 450 bp and two end fragments of 150 bp. In the absence of energy-dependent branch migration, each stable Holliday junction contains one hairpin arm with an *Ava*II site and a sond hairpin arm with a *Bam*HI site. Cleavage at these sites leads to two approx 225 bp hairpin fragments and, following spontaneous branch migration of the second fragment, the same two approx 150 bp end fragments. Thus for Tube 6 and Tubes1–3, bands of 450 bp, 225 bp, and 125 bp should be observed with stoichiometry 1:2:4. In the course of energy-dependent branch migration, the stable Holliday junctions are converted

into heteroduplex reactants again refractory to cleavage by the enzymes. In Tubes 4–5, the appearance of the heteroduplex reactant band running slower than approx 750 bp should be accompanied by disappearance of the hairpin approx 225 bp band and a two-fold decrease in the 150 end-fragment band. When the reaction goes to completion, bands of >750 bp, 450 bp, and 125 bp should be observed with stoichiometry 1:1:2.

9. This resolution assay may be used to determine the time-course of the reaction. The function of native and mutant RuvC proteins from any species may be examined as a function of temperature and solvent. In addition, the combined activities of RuvA, RuvB, and RuvC may be examined. The substrate described here may be modified to examine a variety of potential cleavage sites. The labeled arms may be examined separately by cleavage with a restriction exonuclease with a recognition sequence found only in one of the unique ends. Analysis of the reaction products has shown that the preferred sequence for cleavage by the *E. coli* RuvC is conserved in *T. maritima* RuvC (Gonzalez, S., Rosenfeld, A., Szeto, D., et al, manuscript in preparation). The sequence specificity for other RuvC resolvases is unknown.

References

1. Kowalczykowski, S. C., Dixon, D. A., Eggleston, A. K., et al. (1994). Biochemistry of homologous recombination in *Escherichia coli. Microbiol. Rev.* **58,** 401–465.
2. West, S. (1997) Processing of recombinational intermediates by the RuvABC proteins. *Annu. Rev. Genet.* **31,** 213–244.
3. Iype, L. E., Wood, E. A., Inman, R. B., and Cox, M. M. (1994) RuvA and RuvB proteins facilitate the bypass of heterologous DNA insertions during RecA protein-mediated DNA strand exchange. *J. Biol. Chem.* **269,** 24,967–24,978.
4. Rafferty, J. B., Sedelnikova, S. E., Hargreaves, D., et al. (1996) Crystal structure of DNA recombination protein RuvA and a model for its binding to the Holliday junction. *Science* **274,** 415–421.
5. Hargreaves D., Rice, D. W., Sedelnikova, S. E., et al. (1998) Crystal structure of *E. coli* RuvA with bound DNA Holliday junction at 6 A resolution. *Nat. Struct. Biol.* **5**, 441-446.
6. Iwasaki, H., Shiba, T., Makino, K., et al. (1989) Overproduction, purification, and ATPase activity of the *Escherichia coli* RuvB protein involved in DNA repair. *J. Bacteriol.* **171,** 5276–5280.
7. Mitchell, A. H. and West, S. C. (1994) Hexameric rings of *Escherichia coli* RuvB protein. Cooperative assembly, processivity and ATPase activity. *J. Mol. Biol.* **243,** 208–215.
8. Yu, X., West, S. C., and Egelman, E. H. (1997) Structure and subunit composition of the RuvAB-Holliday junction complex. *J. Mol. Biol.* **266,** 217–222.
9. Ariyoshi, M., Vassylyev, D.G., Iwasaki, H., et al. (1994) Atomic structure of the RuvC resolvase: a Holliday junction-specific endonuclease from *E. coli. Cell* **78,** 1063–1072.

10. Shah, R., Bennett, R. J., and West, S. C. (1994). Genetic recombination in *E. coli*: RuvC protein cleaves Holliday junctions at resolution hotspots in vitro. *Cell* **79,** 853–864.
11. Eggleston, A. K., Mitchell, A. H., and West, S.C. (1997) In vitro reconstitution of the late steps of genetic recombination in *E. coli. Cell* **89,** 607–617.
12. van Gool, A. J., Shah, R., Mezard, C., and West, S. C. (1998) Functional interactions between the Holliday junction resolvase and the branch migration motor of *Escherichia coli. EMBO J.* **17,** 1838–1845.
13. Zerbid, D., Mezard, C., George, H., and West, S. C. (1998) Coordinate actions of RuvABC in Holliday junction processing. *J. Mol. Biol.* **281,** 621–630.
14. Wetmur, J. G., Wong, D. M., Ortiz, B., et al. (1994) Cloning, sequencing, and expression of RecA proteins from three distantly related thermophilic eubacteria. *J. Biol. Chem.* **269,** 25928–25935.
15. Tong, J. and Wetmur, J. G. (1996) Cloning, sequencing, expression, and characterization of RuvB proteins from two distantly related thermophilic eubacteria. *J. Bacteriol.* **178,** 2695–2700.
16. Nelson, K. E., Clayton, R. A., Gill, S. R., et al. (1999) Evidence for lateral gene transfer between Archaea and bacteria from genome sequence of *Thermotoga maritima. Nature* **399,** 323–329.
17. Sambrook, J., Fritsch, E. F., and Maniatis, T. (eds.) (1989) *Molecular Cloning, a Laboratory Manual*, 2nd ed. Cold Spring Harbor University Press, Cold Spring Harbor, NY.
18. Saiki, R. K., Gelfand, D. H., Stoffel, S., et al. (1988) Primer-directed enzymatic amplification of DNA with a thermostable DNA polymerase. *Science* **239,** 487–491
19. Dunderdale, H. J., Sharples, G. J., Lloyd, R. G., and West, S. C. (1994) Cloning, overexpression, purification, and characterization of the *Escherichia coli* RuvC Holliday junction resolvase. *J. Biol. Chem* .**269,** 5187–5194.
20. Tsaneva, I. R., Müller, B., and West, S. C., (1992) ATP-dependent branch migration of Holliday junctions promoted by the RuvA and RuvB proteins of *E. coli. Cell* **69,** 1171–1180.

10

MutS–DNA Interactions and DNase Protection Analysis with Surface Plasmon Resonance

Peter Brooks

1. Introduction

1.1. Protein–DNA Interactions in Mismatch Repair

Elucidation of the mechanisms of recognition of mispaired DNA by MutS and subsequent repair by the multicomponent mismatch repair machinery presents a fascinating challenge *(1)*. ATP binding or hydrolysis are implicated in several steps of the process, including enabling MutS to dissociate from the mismatch, probably through a conformational change *(2)* and for MutSL–DNA interactions *(3,4)*. In addition to mismatch recognition, the protein has significant affinity for normal DNA. This may be important for the protein to scan DNA in search of a mismatch and for its proposed translocation activity. It can be expected, and especially once a MutS-DNA complex structure is solved, that numerous mutants will be generated to study the recognition mechanism. To examine stable and transient DNA interactions of the native and mutant proteins in various conformational states, assays that enable detailed kinetics are desirable. Instrumentation providing such capability is a worthwhile complement to classic assays, such as filter binding and electrophoretic mobility shift assays. Here, we describe techniques for studying the interaction of MutS and MutL with immobilized DNA using surface plasmon resonance and the BIAcore instrument. Such techniques should be readily adaptable to studies of other mismatch binding proteins (MMBP) or other DNA damage recognition proteins.

1.2. BIAcore Technology

In the nanofluidic cartridge of the BIAcore, running buffer is continuously pumped through four 70 nl "flow cells" (FC), each with a ligand (DNA for the

From: *Methods in Molecular Biology, vol. 152: DNA Repair Protocols: Prokaryotic Systems*
Edited by: P. Vaughan © Humana Press Inc., Totowa, NJ

experiments described here) immobilized in a polymer matrix with a surface of 1.2 mm^2 but a depth of only a few hundred nanometers. Sample handling is conceptually similar to a chromatographic system with an injection loop and control valves to direct flow to the cells, singly or in combination, and then to a recovery cup or to waste. Two "pumps" consist of glass syringes with motorized pistons. In the sample compartment, a stainless-steel needle shaft on a robotic arm is programmed to prepare dilutions and inject binding proteins over the surfaces and solutions for washing or surface regeneration. Various tube sizes and 96-well microtiter plates can be used. Automation of transfer, mixing, and injection reduces errors and assures reproducibility. Although a fundamental disadvantage of the instrument is that many samples cannot be analyzed in parallel, the programming permits numerous automated serial analyses for typical cycle times if the reagents remain stable for the series. Samples can be injected to flow over the DNA surfaces for times ranging from seconds to hours.

As the immobilized DNA captures protein, the resulting increase in mass in the polymer matrix produces a slight increase in the refractive index, which is detected by the optical system located behind the derivatized surface. The detection system exploits the optical phenomenon known as surface plasmon resonance (SPR); the BIAcore technology and detailed descriptions of SPR have been reviewed *(5–7)*.

As neither the DNA nor the protein need be labeled, the machine can be considered as an expensive refractometer or as an expensive but exquisitely sensitive analytical balance. The resonance signal is expressed as "response units" (RU) and some tens of RU can be reproducibly measured. Based on the analysis of various immobilized radiolabeled proteins, an increase in 1000 RU is considered to represent 1 ng of protein captured in the surface *(8)*. For nucleic acids, the relationship is 1000 RU for 0.8 ng *(9,10)*. The relation of these two values has apparently not been contradicted by studies where protein/DNA ratios were evaluated by SPR and by an independent technique. For example, a MutS-DNA complex contains two monomers of MutS as determined by SPR and by electron microscopy *(2)*.

2. Materials

2.1. Instrumentation and Sensor Chips

The BIAcore 2000 or BIAcore 3000 (BIAcore, Uppsala, Sweden) permit simultaneous analysis of four FCs and include an integrated sample chamber and robotics for programmed sample injection (*see* **Note 1**). The user can manually execute commands, including establishing a "command queue." Alternatively, the control program includes a simple program language for writing "methods," the series of commands for transfer and mixing, injections, washes, and management of data recording. Methods can include for/end logical loops with variables for repeating the same series of commands for different samples, volumes, and so on. DNA is immobilized, and then protein is injected for a specified association time. Hence, the "incubation" time is determined by the flow rate (typically about 20 µL/min) and the volume injected. At the end of the association phase, running buffer returns to flow over the surfaces for as long as necessary to collect dissociation phase data. Any protein remaining bound to the DNA is then removed with a regeneration solution to enable further injection of protein over the same DNA surfaces. A separate program "BIAevaluation" includes modules for analysis of the kinetograms (*see* **Note 2**). Experiments can be performed between 4°C and 40°C and the metal sample block temperature is controlled through connection to an external recirculating water bath. We keep the blocks containing diluted protein samples at 8°C and perform experiments between 25°C and 37°C. The routine maintenance washes recommended by BIAcore should be respected (*see* **Note 3**). "Sensor chips," type SA (streptavidin immobilized on a carboxymethyl dextran surface), as yet only available from BIAcore, should be prewashed as indicated in the packing notice. To minimize variations in the baseline response, it is wise to first serially inject the various buffers that will be used for DNA immobilization and protein injections including regeneration buffer. A "report point" function includes an evaluation of any baseline drift as a slope of the signal vs time; before DNA or protein injections, we routinely obtain baselines with variations of less than 0.5 RU/min.

2.2. Solutions

As the fluid channels and chambers are of capillary dimensions, it is important to filter buffers and if necessary, protein solutions. Sufficient degassing generally is effected during vacuum filtration of buffers.

1. HBS buffer: 10 m*M* Na-HEPES, pH 7.4, 0.15 *M* NaCl, 3 m*M* ethylenediaminetetraacetic acid (EDTA), 0.005% surfactant P-20, *see* **Note 4**. This buffer is supplied filtered and degassed in convenient packets by BIAcore and is used to continually bathe the cells and channels between experiments.
2. High salt buffer for DNA immobilization: 10 m*M* Na-HEPES, pH 7.5, 0.63 *M* NaCl, 3 m*M* EDTA, 2% glycerol, and 0.005% Tween-20.
3. Running buffer, HAMG: 30 m*M* Na-HEPES, pH 7.5, 220 m*M* NaOAc, 4 m*M* $MgCl_2$, 0.1 m*M* EDTA, 2% glycerol, and 0.005% Tween-20.
4. Concentrated protein dilution buffer, HABG: HAMG without $MgCl_2$ and with 5 m*M* β-mercaptoethanol. Depending on the extent of protein dilution from a 50% glycerol stock, the glycerol concentration in this buffer can be reduced (*see* **Subheading 3.1.**).

5. Reaction buffer, HABMG: HAMG with 5 m*M* β-mercaptoethanol. This is the same as "sample buffer" because the preinjection protein diluent dictates the composition of the incubation for the protein-DNA interaction.
6. Surface regeneration buffer for protein removal: HAMG containing 0.04% sodium dodecyl sulfate (SDS).
7. Surface regeneration buffer for DNA removal: DNase I is diluted in HABMGC (HABMG with 0.5 m*M* $CaCl_2$) to 3 μg/mL.

2.3. MutS and MutL Proteins

The protein preparations used (GeneGuard-S and GeneGuard-L, Genoscope, Evry, France) were found to be free of nuclease and protease for the immobilized DNA experiments described here. Highly concentrated proteins stored in a 50% glycerol solution are manually diluted with HABG to a concentration about 10 times higher than the final experimental concentrations. MutS, 1.2 μ*M*, is stable at least 30 h at 8°C in HABG.

2.4. DNA Substrates

1. Use standard annealing methods with heat denaturation to form heteroduplex or homoduplex oligonucleotide substrates. Use of an excess of a nonbiotinylated oligo minimizes immobilization of the complementary single-stranded biotinylated oligo on the SA surfaces (*see* **Note 5**).
2. Long substrates composed of sequentially overlapping oligonucleotides without ligation are adequate for certain studies of MutSL–DNA interactions *(4)*.
3. Amplicons produced by PCR from a reference sequence using one biotinylated primer are annealed (using alkaline denaturation) with a nonbiotinylated amplicon from a mutated sequence either in equimolar amounts or with an excess of the nonbiotinylated DNA to increase the proportion of heterohybrid DNA immobilized.

3. Methods

New users should obtain instruction from experienced users for method programming and on the basic handling of the instrument and sensor chips. The instrument and applications manuals provided by BIAcore are sufficiently comprehensive for initial experiments and some excellent practical advice and admonishments have been presented in a previous volume in this series *(11)*.

3.1. Management of "Bulk Effects"

As the response signal is extremely sensitive to changes in refractive index, even slight differences in composition between running buffer and the injected sample buffer can generate immediate and large changes in response signals, the "bulk effect." In many situations, this is tolerable, especially through subtraction of the kinetogram of a FC without DNA (*see* **Subheading 3.2.**), but if accurate initial association rates are to be measured, such changes can obscure the initial increase in RU that is due to protein binding. In addition, for real-time monitoring of a manual injection of DNA, it may be difficult to establish the initial RU level taken as the zero level. Thus, it is preferable that the refractive indices of the running buffer and injected sample differ as little as possible.

1. As the MutS-DNA interaction is little affected by glycerol concentrations ranging from 0% to 20%, the glycerol in the protein dilution buffer can be varied slightly around 2% to compensate for refractive index differences between the injected sample and the running buffer (*see* **Note 6**).
2. Alternatively, use of the «co-inject» command enables an injection of the sample buffer for a specified time and then an immediate injection of the sample. Hence, if the original protein solution has been sufficiently diluted into sample buffer, then bulk effects are minimized at the beginning of the association phase. An additional advantage of this procedure is that if any component of the sample buffer perturbs the conformation of the dextran surface or the immobilized DNA, the surface will have the opportunity to stabilize in the sample buffer before the sample is injected. An example is shown in **Subheading 3.5.1.**
3. If the protein–DNA interaction is compatible with running buffer, then desalting chromatography could be used to replace the protein buffer with running buffer. However, for experiments with many sequential cycles requiring keeping a protein stock for many hours in a sample block, the running buffer may not be the best for protein stability. We prefer to continually bathe the DNA surfaces with a mercaptan-free, Mg-containing buffer (HAMG), but to avoid hours-long exposure of the proteins to metals, we use the metal-free buffer, HABG, to dilute the stock protein solution.

3.2. Immobilization of DNA

Immobilization can be performed using a programmed method or semimanually. With the latest version of the control program, the user can specify the number of RU to be immobilized. This is only effective if the refractive indices of the running buffer and the DNA buffer are the same. The "manual inject" command is used to precisely control the amount of DNA immobilized.

1. It is best to temporarily use the high salt buffer as running buffer to eliminate large bulk effects during the DNA injection. After establishing a stable baseline response, dilute biotinylated hybrid double-stranded DNA with high salt buffer and inject over one SA FC until the desired RU is obtained. For a typical experiment, FC1 has no DNA, FC2, homoduplex, and FCs 3 and 4, two different mismatched DNAs (*see* **Note 5**).
2. For accurate kinetic evaluation, use the minimum amount of DNA to obtain reliable values when protein is injected. With increasing levels of immobilized DNA, "mass-transport" effects become problematic for kinetic analysis (*7*; *see* BIAcore manuals). For example, when 100 pg of a 25 bp duplex (125 RU) is immobilized in the FC surface matrix (volume of about 120 pL), the DNA concentration in the

matrix is about 50 μ*M*. Thus, at a time during dissociation when a significant fraction of the DNA is unbound, protein leaving a DNA molecule will readily encounter and rebind another substrate molecule before it is able to diffuse away from the surface. Hence, the observed k_{off} is slower than the true k_{off} of the protein–DNA interaction. Mass-transport effects also increase the observed k_{on}. Increasing the flow rate has only a marginal effect and so reduction of the surface concentration is the best means to alleviate the problem. (An exception to this guideline is discussed in **Subheading 3.5.2.**)

3. Short oligonucleotide duplexes are readily immobilized, such that to control the rate of immobilization, especially to avoid overshooting the amount necessary, the solution must be diluted to about 10 n*M*. However, with substrates over 100 bp, it is more difficult to achieve the desired level of immobilization (*see* **Note 7**), and so to economize the quantity needed for the injection, the flow rate is reduced to a few μL/min.
4. Chips can be "undocked" using the option for drying the surfaces and then stored in 50-mL capped tubes at 4°C for weeks before reuse with reasonably reproducible results. Although the chips are relatively expensive, if the injected analytes are free of protease, nuclease, and helicase, the DNA surfaces can be regenerated by complete protein removal and used for dozens of experiments. In addition, when there is some loss of DNA either by contaminating nucleases, or during nuclease protection experiments, surfaces can be treated with excess DNase and then biotinylated DNA can be reimmobilized many times before the SA begins to be saturated.

3.3. Protein Injection

MutS does not require a metal cofactor for binding to DNA and so a Mg-free buffer with EDTA can be used, especially if contaminating nucleases may be present (e.g., *see* **Subheading 3.5.2.**). Nevertheless, we routinely include $MgCl_2$ to study both MutS-DNA and ATP-dependent MutSL–DNA interactions.

1. As concentrated MutS tends to oligomerize, even in 50% glycerol solutions stored at –20°C, we mix by gentle pipeting before removing an aliquot. Typically, a 1.2-μ*M* solution is prepared in HABG and kept in an Eppendorf tube in the cooled sample block. To minimize evaporation, the tube can be covered with parafilm, which the sample needle can puncture.
2. For each cycle, an empty BIAcore narrow polypropylene tube with its piercable cap is used for automated sample preparation. Immediately before injection, the program directs transfer of reaction buffer, MutS, MutL, and nucleotides, and mixing in the tube. For a standard injection, the final volume should exceed the injection volume by 40 μL, as 30 μL is lost in the sample loop and although the injection needle plunges to nearly the conical bottoms of the sample tubes, sufficient volume should remain to avoid injection of air bubbles (*see* **Note 8**).
3. Typically, we inject MutS between 10 and 200 n*M* for 5 to 20 min, depending on the amount of DNA immobilized and the kinetic information sought. A flow rate

of 10 µL/min is suitable, but for maximum economy with precious protein samples, flow can be reduced to 1 µL/min to increase contact time for a given volume injected.

4. Before launching a lengthy series of programmed injections, it is wise to monitor two or three cycles, including an injection of buffer without protein and two identical protein injections to ensure that the FCs are behaving consistently and that the programmed method performs correctly. Although the software will detect and flag syntactical and logical errors in a program, it cannot know if programmed sample addresses, times, and volumes are incorrect.

3.4. Regeneration by Removal of Protein or DNA

1. After the dissociation phase, the protein is easily removed by two sequential 30 sec injections of running buffer containing SDS.
2. Other regenerants added to HAMG such as 4 m*M* ATP, 4 *M* NaCl, or 0.4 *M* sodium phosphate cause rapid dissociation of the complex, but each leaves some residual protein that must be removed with SDS.
3. For near complete degradation of the immobilized DNA, inject DNase I for 10 min followed by protein regeneration buffer. Then reimmobilize DNA. Neither MutS nor MutL are captured by the residual oligonucleotides (average size of 6 nucleotides). (For an example, *see* **Subheading 3.5.6.**).

3.5. Experimental Applications and Data Analysis

3.5.1. Treatment of Raw Data and Comparative Kinetics

A typical experiment with duplex oligonucleotides of 39 bp is shown in **Fig. 1**. The three panels show the appearance of the kinetograms before and after setting a common zero level and correction with the kinetogram from the surface without DNA. In this experiment, the buffer HABMG was replaced with a buffer that differed from the running buffer by about 500 RU. To pre-equilibrate the immobilized DNA in this buffer and to avoid a large bulk effect at the beginning of association, we used the co-inject command to wash the surface with the buffer for 3 min before protein injection. As the binding to homoduplex DNA is not negligible and because we wish to explore factors that influence this binding, we generally do not correct the heteroduplex binding for the non-mismatch specific binding.

The absolute rates of association (k_{on}) and dissociation (k_{off}) for MutS–DNA binding have been difficult to reliably assess because of the formation of several types of complexes and difficulties caused by mass transport effects. Nevertheless, comparison of maximum binding levels achieved and visual comparison of kinetograms are instructive. For example, in some conditions, the association phase with the IDL-1 substrate of **Fig. 1** is sigmoidal, indicating a cooperative association, and a suggestion that binding mechanisms with IDL substrates and single base mispairs may differ. Different binding conditions

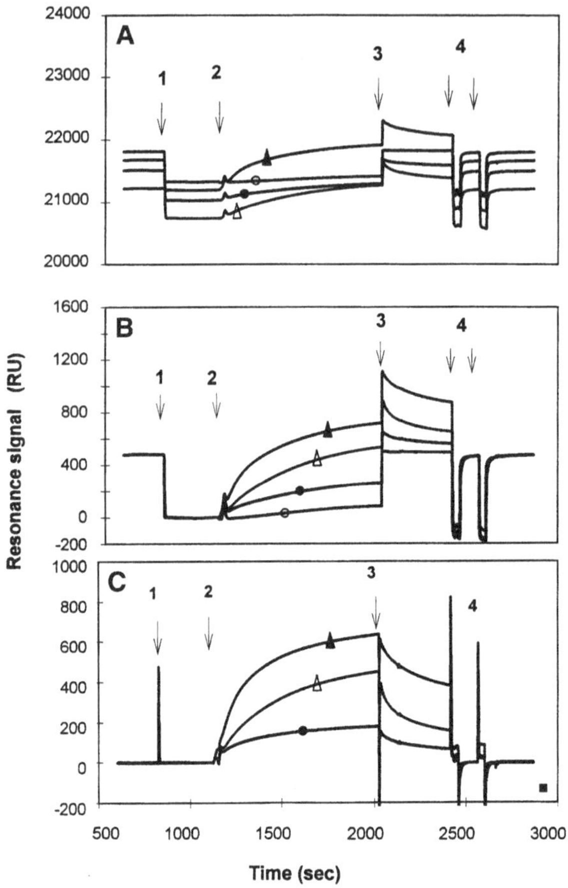

Fig. 1. Kinetics of MutS interaction with duplex oligonucleotide substrates: processing of BIAcore data. The kinetograms shown are for FCs with no DNA (open circles), or with 200 RU of immobilized homoduplex (filled circles), or T-G (filled triangles) and IDL-1 (open triangles; insertion/deletion of one nucleotide) mispaired oligonucleotides. The arrows indicate (1) injection of diluent to equilibrate DNA, (2) injection of MutS in diluent, (3) end of protein injection—return to running buffer and beginning of dissociation phase, and (4) elution of remaining protein with SDS and return to baseline DNA level. (**A**) Unprocessed data. Initial RU differences are due to different amounts of carboxymethyl dextran and streptavidin. (**B**) A baseline zero level is defined at point 2 just before injection of protein. The rapid changes in signal at points 1, 3, and 4 are caused by bulk effects. (**C**) The data from the surface with no DNA are subtracted from the three experimental kinetograms. This removes the bulk effects and corrects for any nonspecific binding of the protein to the SA-dextran matrix. The flow cells are arranged in series in the flow channel; hence, the nonsynchronous passage of buffer fronts over the four surfaces results in the appearance of artifactual spikes when the control kinetogram is subtracted.

can generate complexes that decay during dissociation with differing rates. Semiquantitative comparisons for such data can simply be expressed as the time resulting in decay of the RU level to half its maximum at the end of the association phase. A programmed method can specify report points for recording RU levels, for example, 2 min after injection, 30 s before the end of association, and 2 min and 10 min into the dissociation phase. The values from multiple cycles are grouped in one report table for manipulation and plotting. Thus, differences in the data between cycles can be rapidly evaluated without the tedium of examining each set of kinetograms (for an example, *see* **Subheading 3.5.3.**).

3.5.2. Monitoring MutS Purification

An exception to the general policy of minimizing the amount of immobilized DNA arises when MutS concentrations are measured. When the surface has a high concentration of DNA substrate sites, a protein molecule that has entered the surface matrix is certain to bind a DNA molecule before it diffuses back into the flow channel. Thus, the apparent rate of binding is limited by the rate of diffusion to the surface (mass transport) and, hence, the initial rate of binding observed is proportional to the amount of protein injected. BIAcore initial rate data have been used to follow the purification of a ligand specific for an immobilized receptor ***(12)***. With a SA surface saturated with biotinylated heteroduplex DNA, the binding rate increased linearly with increasing MutS concentrations over a 500-fold range (**Fig. 2, A,B**). Thus, chromatographic fractions were rapidly assessed for binding activity as brief injections permitted collection of sufficient initial rate data (**Fig. 2, C,D**). In addition, because only a small fraction of the DNA is bound at each injection, sequential injections are possible on the same surface without intermediate regeneration. The ratio of the concentration of the binding activity to the protein concentration varied less than 15% for the peak fractions, indicating that the peak represented a highly purified fraction (data not shown). If flow rates were reduced to increase contact times, as little as 10 ng could be used to obtain a significant binding rate (*see* **Note 9**).

3.5.3. Comparison of Binding Conditions

Comparison of the protein–DNA interactions in various solution conditions is useful for several purposes and is readily achieved with the BIAcore. For eventual use of MutS for mutation detection, it is desirable to optimize the discrimination ratio, namely the ratio of mismatch-specific to nonspecific binding. On the other hand, to evaluate the effects of various mutations on the different types of MutS-DNA interactions, we chose to find conditions that permitted an adequate discrimination ratio while preserving substantial nonspecific binding.

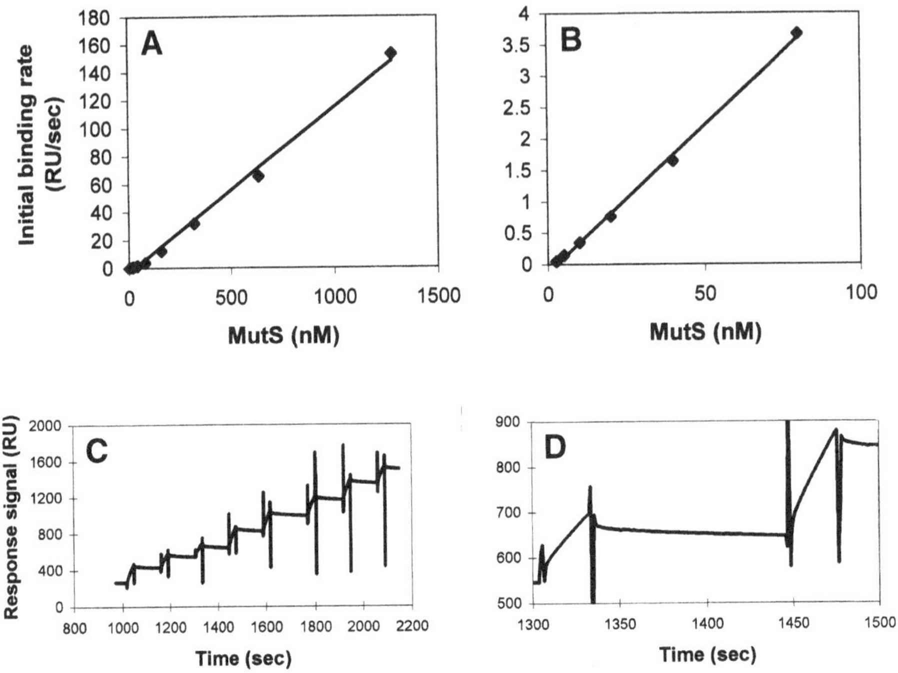

Fig. 2. Determination of MutS concentration. **(A) and (B)**. Initial binding rates are plotted as a function of MutS concentration. A biotinylated heteroduplex oligonucleotide duplex was injected until the SA surface was saturated (about 2000 RU). MutS was injected and binding rates were obtained from the initial linear association phases of the kinetograms. Rates were determined from the first few min for the most dilute concentrations and from the first 10–30 s for the most concentrated. (Data for the lowest concentrations in **A** are shown enlarged in **B**.) **(C) and (D)**. Serial injection of chromatographic fractions. $MgCl_2$ was replaced with EDTA in the buffers to minimize any nuclease activity in the column fractions. Data are shown for 8 serial 30 s injections of fractions that included a MutS elution peak. The method program included dilution of the fractions to similar protein concentrations, as guided by the A_{280} profile of the elution. Hence the slopes shown vary within a limited range. **D** is an enlargement of two of the injections to show the linearity of the initial binding response.

Fig. 3 shows an example of such a screening experiment. As the duration of the series was more than 20 h, the final cycle was programmed to repeat the first cycle (*see* **Note 10**). Different conditions enhanced (cycles 4 and 7) or abolished binding specificity (cycle 6) and, at least in one case, certain conditions enabled preferential binding to the T-G mismatch as compared with the IDL-1 (cycle 4). In addition, examination of the kinetograms revealed interesting differences in association and dissociation behaviors (not shown). Based

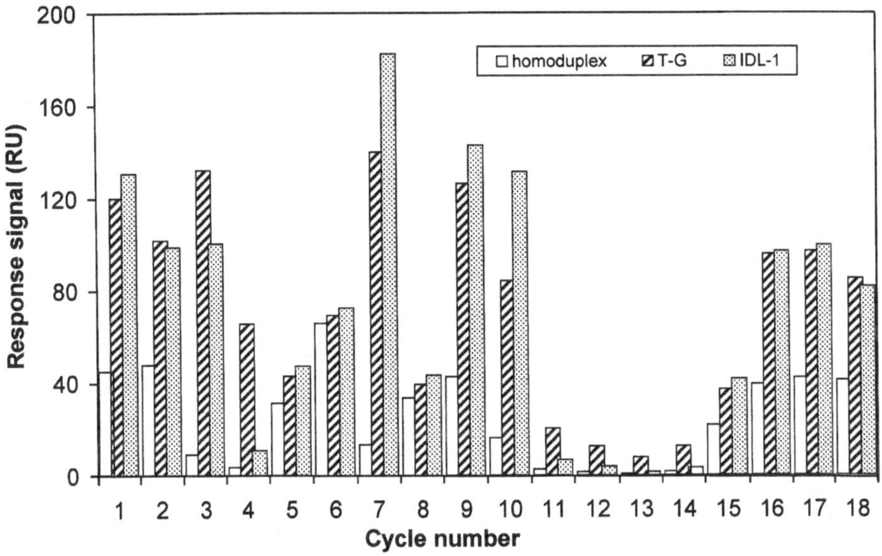

Fig. 3. Comparison of binding conditions. MutS was diluted into 17 different reaction cocktails and injected over the surfaces with mismatched and homoduplex oligonucleotide substrates. Amplitudes shortly before the end of the association phase are plotted. Cycles 1 and 18 are with identical conditions. In this experiment, the thermoblock temperature was not controled and so the MutS solution endured 25–30°C for several hours in the sample compartment.

on the results of similar experiments, we selected HABMG as a standard reaction buffer that permits monitoring of specific and nonspecific interactions and contains a physiological concentration of acetate.

3.5.4. Detection of Transient Interactions of MutS and MutL with DNA

Interactions of proteins with DNA that involve rapid turnover of the proteins may be difficult to detect with some classic techniques. Such interactions could be missed entirely using gel filtration or gel retardation where unbound protein is separated from the unstable nucleoprotein complex. Alternatively, with nuclease footprint experiments, in which free protein is generally present and active during the nuclease challenge, a dynamic interaction might not be distinguished from a stable complex. In a BIAcore experiment, data from both conditions, namely association with excess free protein and evaluation of the stability of the complex in the absence of unbound protein, are observed in the same kinetogram. Hence, dynamic interactions are immediately apparent. For example, when MutS and MutL with ATP were injected over DNA surfaces, the response signal rapidly rose to a substantially higher level than when MutS alone was injected, but decayed very rapidly. As complexes formed during the association phase are probably susceptible to the same rapid departure of protein, it was deduced that the strong association response, both in rate and amplitude, involved a rapid association of the proteins with DNA concurrent with a rapid dissociation *(4)*.

3.5.5. Protein Recovery

A recovery cup permits collection of most of the injectate so that much of the DNA or protein not bound during the injection can be collected for subsequent injections. As our reagents are not limiting, we have not exploited this feature. BIAcore has suggested that protein could also be collected during dissociation or by elution with the regeneration solution. However, we found that the recovery function was not useful for analyzing MutS and MutL eluted with SDS. It was not possible to completely wash the walls of the binding chamber free of nonspecifically bound protein and so protein was recovered even from the cell with no DNA (*see* **Note 11**). However, guided by the results from BIAcore experiments, we readily mimicked the experiments with DNA immobilized on magnetic beads to show that MutS and MutL were present in the protein-DNA complexes *(4)*.

3.5.6. Nuclease Protection

In addition to the evaluation of binding interactions, enzymatic reactions can be performed with the immobilized DNA as substrate. For example, SPR has been used to monitor immobilized template-directed nucleic acid synthesis by polymerases *(10)*. To mimic a classic footprint experiment, we used the co-inject command to first establish the ATP-dependent interaction with MutS and MutL and then immediately injected the same solution with DNase I (*see* **Note 12**). The capability of the BIAcore to accurately measure the remaining DNA enabled the conclusion that the dynamic MutSL–DNA interaction offered more protection to the nuclease than that provided by MutS alone *(4)*. If we allowed a partial dissociation to occur before injection of DNase, the protection was the same as that provided by MutS, even though MutS and MutL were still present. Thus, we concluded that the enhanced protection was dependent on the presence of the exchangeable unbound proteins *(4)*. An analogous experiment with classic footprinting would require purification of the nucleoprotein complex to remove unbound protein before the DNase challenge.

4. Notes

1. BIAcore 1000 and BIAcore-X are simpler versions but lack essential features required for the experiments described here. The BIAcore-X requires manual injections and the BIAcore 1000 requires separate injections for control cells and each cell with DNA.

2. BIAcore uses the term "sensorgram" for the plots of response signal (RU) vs time. "Kinetogram," a general term for any signal varying as a function of time, is suggested for data generated by instruments that continuously monitor reaction kinetics, in analogy to spectrogram, a plot of a signal varying in response to varying wavelengths.
3. As the machine represents an important investment, in many research institutions it is available in a core facility which is responsible for the simple but regular maintenance essential for reliable functioning. Alternatively, a policy, such as at Genoscope, permits experiments by extramural users when scheduling permits. As the physical experimental setup at the instrument is straightforward, it is conceivable that such use could be from a distance, with posting of ligands and analytes and electronic exchange of methods and results.
4. A 10% solution of "surfactant P20" is furnished by BIAcore and can be substituted by Tween-20. Use of such a nonionic detergent is highly recommended to reduce non-specific interactions in the fluid channels and FCs. The detergent apparently interacts with the surface, as injection of a buffer without detergent requires many minutes of washing before the response stabilizes. If an injection sample does not contain the detergent, either use a detergent-free running buffer or to obtain accurate data at least for the association phase, use the co-inject command to pre-inject detergent-free diluent as mentioned in **Subheading 3.1.**
5. In principle, it should be possible to first inject only the biotinylated oligo and then its non-biotinylated complement, but in practice, this yields only about an 80% efficiency of conversion to duplex DNA. Thus, it is best to preform the duplexes with salts and temperatures adapted for the particular substrates.
6. Adjustment of the glycerol concentration to eliminate bulk effects can be used even though other components, such as salts, might be the cause of the refractive index difference.
7. This may be because of the negative surface charge of the carboxymethyl dextran matrix as the derivitization procedures with SA are never complete.
8. Occasionally, the mixing will introduce some bubbles due to detergent; if this is a consistent problem, increase the final volume.
9. Because the binding is diffusion-limited, the difference in binding between heteroduplex and homoduplex is less pronounced than with lower density surfaces yet is sufficient to identify a specific MMBP.
10. The apparent loss of activity (compare the duplicate cycles 1 and 18), partially caused by lack of sample block cooling, did not affect the qualitative interpretation of this type of experiment. Excellent quantitative reproducibility is achieved if the protein solution is kept in chilled sample blocks.
11. Protein that does not contribute to the response signal may nevertheless be nonspecifically adhering to the remaining surfaces of the chamber or sequestered at some point in the flow cartridge passages, and although not cleared by running buffer, is washed off by the SDS elution buffer. BIAcore claims that the BIAcore 3000 includes modifications that have addressed this problem.

12. These experiments involved immobilization of about 100 RU of 149 bp substrates, regeneration with excess DNase after each cycle, and then reimmobilization of DNA on the same sensor chip. As each cycle of DNA immobilization exploited only a small fraction of the SA capacity, the surfaces could be regenerated by DNase and rederivatized with DNA for at least 40 cycles.

Acknowledgments

The author would like to thank Jean Weissenbach for helpful encouragement, Nathalie Surin and Charles Marcaillou for expert technical assistance, and the Association Française contre le Myopathie for support during the preliminary phase of this work at the Généthon laboratories. P.B. is on leave to Genoscope from the Centre National de la Recherche Scientifique.

References

1. Modrich, P. (1991) Mechanisms and biological effects of mismatch repair. *Annu. Rev Genet.* **25,** 229–53.
2. Allen, D. J., Makhov, A., Grilley, M., et al. (1997) MutS mediates heteroduplex loop formation by a translocation mechanism. *EMBO J.* **16,** 4467–4476.
3. Grilley, M., Welsh, K. M., Su, S. S., and Modrich, P. (1989) Isolation and characterization of the Escherichia coli *mutL* gene product. *J. Biol. Chem.* **264,** 1000–1004.
4. Galio, L., Bouquet, C., and Brooks, P. (1999) ATP hydrolysis-dependent formation of a dynamic ternary nucleoprotein complex with MutS and MutL. *Nucleic Acids Res.* **27,** 2325–2331.
5. O'Shannessy, D., Brigham-Burke, M., Soneson, K. K., et al. (1994) Determination of rate and equilibrium binding constants for macromolecular interactions by surface plasmon resonance. *Methods Enzymol.* **240,** 323–349.
6. Karlsson, R. and Stahlberg, R. (1995) Surface plasmon resonance detection and multispot sensing for direct monitoring of interactions involving low-molecular-weight analytes and for determination of low affinities. *Anal. Biochem.* **228,** 274–280.
7. Fivash, M., Towler, E. M., and Fisher, R. J. (1998) BIAcore for macromolecular interaction. *Curr. Opin. Biotechnol.* **9,** 97–101.
8. Sternberg, E., Persson, B., Roos, C., and Urbaniczky, C. (1991) Quantitative determination of surface concentration of protein with surface plasmon resonance by using radiolabelled proteins. *Colloid Interface Sci.* **143,** 513–526.
9. Fisher, R. J., Fivash, M., Casas-Finet, J., et al. (1994) Real time measurements of *Escherichia coli* single-stranded DNA binding protein (SSB) to polydeoxythymidylic acid reveal single-state kinetics with steric cooperativity. *Methods: Comp. Methods Emzymol.* **6,** 121–133.
10. Buckle, M., Williams, R. M., Negroni, M., and Buc, H. (1996) Real time measurements of elongation by a reverse transcriptase using surface plasmon resonance. *Proc. Natl. Acad. Sci. USA* **93,** 889–894.
11. Panayotou, G. (1998) Surface plasmon resonance. Measuring protein interactions in real time. *Methods Mol. Biol.*, **88,** 1–10.
12. Lackmann, M., Bucci, T., Mann, R. J., et al. (1996) Purification of a ligand for the EPH-like receptor HEK using a biosensor-based affinity detection approach. *Proc. Natl. Acad. Sci. USA* **93,** 2523–2527.

PART II

Use of DNA Repair Enzymes as Tools in Molecular Biology

11

Flexible Genetic Engineering Using RecA Protein

Lance J. Ferrin

1. Introduction

1.1. Types of Projects Where RecA Protein Is Useful

The analysis or manipulation of small segments of DNA in plasmids has long been routine, but standard methods fail or are very time consuming when applied to long segments or complex mixtures of DNA. At present, RecA protein-based techniques usually have been used for making defined changes in large DNA constructs or for precise physical mapping of large regions of genomic DNA from several organisms. The techniques offer a rapid and versatile "cut and paste" approach to several other problems in molecular biology, and their use will probably increase as the goals of experiments with DNA and chromosomes evolve in sophistication. For example, they can now be used to analyze rearranged DNA in individual tumors or genetic deletions in patients with inherited or acquired syndromes. A new variation involving sequence-specific ligation may also eventually be used to rapidly screen libraries, label DNA fragments, or amplify large segments of DNA *(1)*.

Specific examples where the techniques have proven especially useful include: mapping genomic DNA regions containing large amounts of repetitive DNA such as telomeres *(2–4)*, mapping yeast artificial chromosomes and the corresponding region in human genomic DNA *(5,6)*, mapping genomic DNA containing CpG islands *(2)*, and creating point mutations in large DNA constructs *(7,8)*. Whereas these have been the most popular applications, a full bibliography of RecA protein technique applications has been recently published *(1)*.

From: *Methods in Molecular Biology, vol. 152: DNA Repair Protocols: Prokaryotic Systems*
Edited by: P. Vaughan © Humana Press Inc., Totowa, NJ

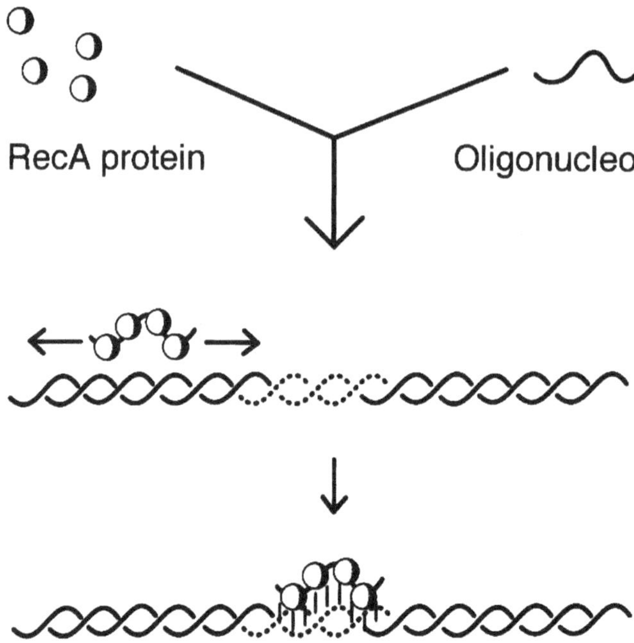

Fig. 1. Formation of a synaptic complex. The dotted region of the duplex is homologous to the oligonucleotide.

1.2. Principles of RecA Protein Enzymology

Genetic recombination in bacteria is a multistep process, and RecA protein plays a complicated role in that process, but the techniques described here only rely on the simple in vitro model pairing reaction shown in **Fig. 1**. First, RecA protein is added to an oligonucleotide and a copolymer is formed. Enough RecA protein is added to completely coat the oligonucleotide, but excess RecA protein is avoided because free protein will inhibit subsequent reactions. Second, duplex target DNA is added and the copolymer forms a complex with the duplex. Although the nature of this nonspecific complex is not well understood, an attractive hypothesis is that the copolymer slides along the duplex with the RecA protein mediating a search for sequence identity between the oligonucleotide and the duplex. Third, where there is complete sequence homology between the oligonucleotide and the duplex, a tight complex (termed a "synaptic complex") forms between the copolymer and the duplex. This complex probably has the structure of a triple helix. Under optimal conditions, this complex is stable for hours, but can be quickly dissociated under mild conditions. The complex blocks the action of several enzymes including restriction endonucleases, DNA methylases, and polymerases, and thus serves as a reversible sequence-specific "masking tape" for DNA. A curious feature of the reaction is that. currently, it can only be carried out to about 80% completion, and depending on the application, products generated by incomplete protection may require an additional step, such as gel purification.

1.3. Protection of a Selected DNA Site from Restriction Endonuclease Cleavage: The Synaptic Complex Reaction

The simplest application using RecA protein is to block a selected site from restriction endonuclease cleavage. This is shown in **Fig. 2A**. An oligonucleotide that has the sequence of the site to be blocked is synthesized, with the restriction endonuclease cleavage sequence generally placed near the middle of the oligonucleotide. A synaptic complex is formed at the site and a restriction endonuclease is added. If the endonuclease has at least some activity in the RecA protein buffer, then cleavage takes place at all sites except for the one blocked by the synaptic complex.

1.4. Cleavage at a Selected DNA Site: RecA-Assisted Restriction Endonuclease (RARE) Cleavage

The scheme for cleaving at a single selected DNA site is shown in **Fig. 2B**. Here, a synaptic complex is used to protect a selected restriction endonuclease site. A methylase that blocks the restriction endonuclease is used to methylate all the unprotected sites. The synaptic complex is removed and the restriction endonuclease is added to cleave at the previously protected site.

1.5. Ligation at a Selected DNA Site: RecA-Assisted Ligation and RecA-Assisted Cloning

This variation is shown in **Fig. 2C**, and is the functional reverse of RARE cleavage in that DNA is joined, instead of cleaved, in a sequence-specific manner. The DNA must first be digested with a restriction endonuclease that creates a 3'-recessed end (most endonucleases are of this type). Synaptic complexes are used to protect the ends of a selected linear fragment of DNA, and DNA polymerase is used to fill in the ends of unprotected fragments. After removal of the synaptic complexes, a second DNA fragment with a compatible cohesive end is added and under specific ligation conditions, ligation to the selected fragment preferentially occurs. Commonly, the second fragment is a linker or a short-labeled duplex, but the only requirements of the second fragment is that it has a cohesive end complementary to at least one end of the first fragment and that it function as a substrate for a DNA ligase. For example, biotinylated duplexes can be used, and one can imagine other interesting substrates, such as hairpins or branched DNA structures. If a vector is ligated to the selected fragment and the product used to transform

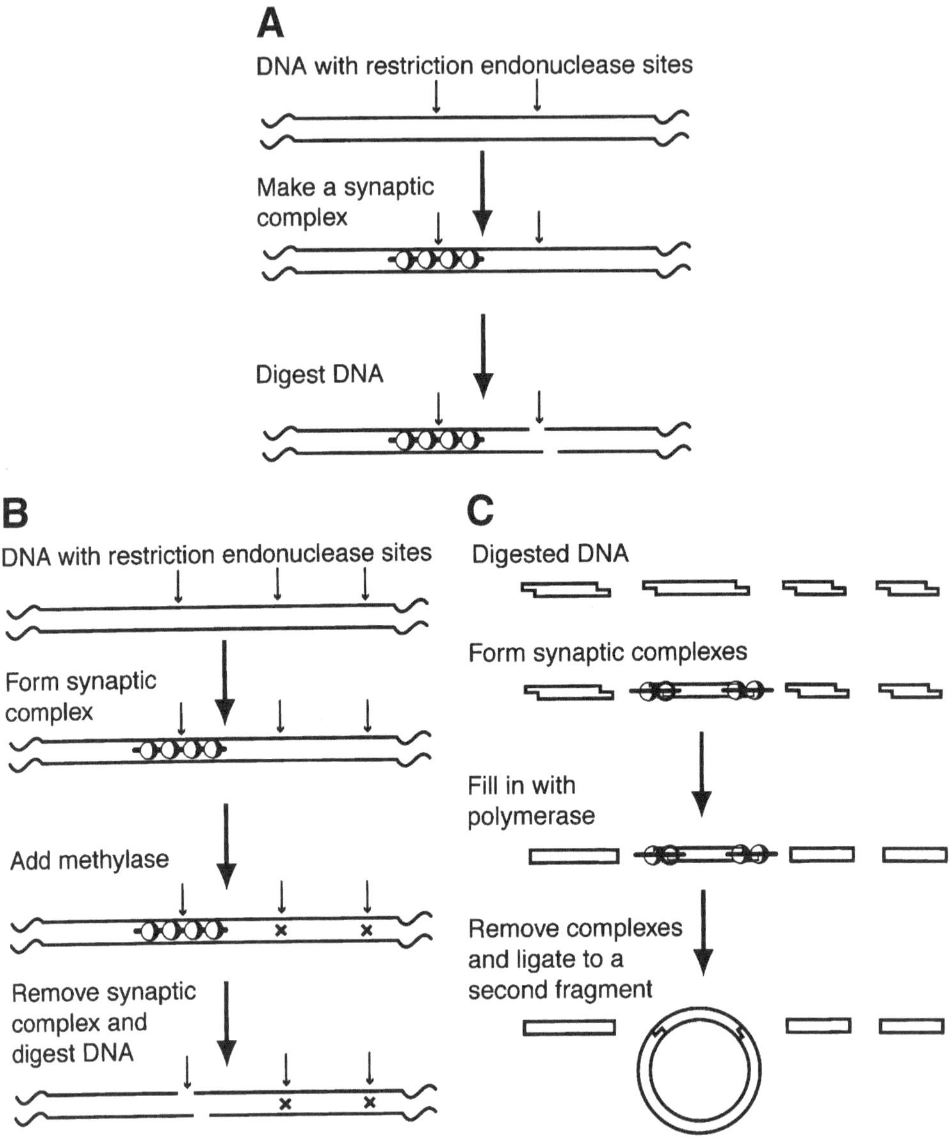

Fig. 2 (**A**) Protection of a selected DNA site from restriction endonuclease cleavage. (**B**) RARE cleavage. The "X" shows sites that are methylated and rendered resistant to restriction endonuclease cleavage. (**C**) RecA-assisted ligation and RecA-assisted cloning. Ligation (with circularization) of a vector to a selected fragment is shown.

bacteria, the selected fragment is preferentially cloned, and when RecA protein is used for this purpose, the variation is termed RecA-assisted cloning. This new technique has sufficient specificity to clone a selected fragment directly from human genomic DNA, but is relative0ly inefficient in terms of the amount of starting material required.

1.6. Construction Strategies

The use of RecA protein greatly extends the capabilities of standard molecular biology techniques, and by combining both, there are very few limitations as to the possible constructs that may be created. Whereas constructs in all standard vectors may be used, RecA-protein techniques offer a unique advantage over standard techniques when creating very large constructs. This has been especially useful with bacterial artificial chromosome (BAC) and P1-derived artificial chromosome (PAC) vectors, which are convenient to use and permit constructs of up to about 300 kb in size.

Because RecA protein and the appropriate oligonucleotide can be used to protect any site, the only sequence limitations are those imposed by the specificity and characteristics of the ancillary enzymes that are used in the procedures. In the case of protection from restriction endonuclease cleavage, the endonuclease chosen must have at least some activity under the conditions optimal for RecA protein. In particular, the RecA protein buffer used has a low ionic strength and the cleavage must be done at 37°C, but almost all endonucleases function under these conditions. In the case of RARE cleavage, the endonuclease used for cleavage must be capable of being blocked by a methylase. Fortunately, there are several methylases that are commercially available and active in the RecA protein buffer, and some block more than one restriction endonuclease. A list of endonucleases and their methylation sensitivity has been published *(9)*. In the case of RecA-assisted ligation and RecA-assisted cloning, the only requirement is that the DNA used must have at least one 3'-recessed end.

When cleaving fragments using RARE cleavage, restriction endonucleases with short recognition sequences allow the most flexibility, but require a high efficiency of methylation, especially when cleaving large fragments. For reason of cost, purity, and methylation activity, *Eco*RI methylase has been used almost exclusively when cleaving fragments greater than 100 kb in size.

Another important feature is that the RecA protein site protection efficiency is at best only about 80%. The biochemical reason for this is unknown, but the practical application is that it is usually best to try to protect no more than two sites in any given reaction. Also, some unwanted products will always be generated, and when working with very large or complicated constructs, it is usually advisable to gel purify or subclone intermediates.

Whereas RecA protein techniques have usually been used to alter cloned DNA constructs, the techniques have sufficient specificity to work well with genomic DNA. Genomic DNA can be mapped by performing RARE cleavage at two defined sites and determining the size of the fragment created, and fragments larger than 1000 kb have been created. RecA-assisted cloning has been used to isolate a specific fragment directly from human DNA, although this technique, when applied to complex genomes, requires a fairly large amount of starting DNA, and has yet to be applied to large segments of DNA.

2. Materials

2.1. Core Reagents Needed for Any RecA Protein Reaction

1. RecA protein buffer (5X stock): 125 m*M* Tris acetate, pH 7.85, 20 m*M* Mg acetate, 2.5 m*M* spermidine trihydrochloride (Sigma) and 2 m*M* dithiothreitol (DTT), (*see* **Note 1**).
2. Adenosine 5'-diphosphate (ADP) and ATPγS (10X stock): 11 m*M* ADP and 3 m*M* ATPγS (Fluka). Determine the concentration of each assuming that a 1 *M* solution has an absorbance at 259 nm of 15,400.
3. Bovine serum albumin (BSA): 2 mg/mL. Use a high-quality BSA, such as Sigma #A 7638.
4. RecA protein. There are several suppliers, but choose a lot where the concentration of protein is at least 4 mg/mL, as the protein is usually supplied as a 50% glycerol solution, and the final concentration of glycerol in the reaction should not exceed 5%.
5. 10% sodium dodecyl sulfate (SDS).
6. Oligonucleotides (*see* **Note 2**).
7. Restriction endonucleases. The ones chosen must have at least some activity in RecA protein buffer (*see* **Note 3**).
8. Starting DNA substrate: concentration should be at least 100 μg/mL.

2.2. RARE Cleavage

Need above reagents plus:

1. 5 m*M* ethylene glycol-*bis* (β-aminoethyl ether) N,N,N',N',N'-tetraacetic acid (EGTA).
2. Methylase (*see* **Note 4**).
3. S-adenosylmethionine: provided with methylases from New England Biolabs. Dilute to 2.4 m*M* immediately prior to use.
4. Restriction endonuclease and buffer (*see* **Note 3**).

2.3. Working with DNA Embedded in Agarose

1. Mammalian cells. If working with bacteria or yeast, similar protocols are available ***(10–12)***.
2. Phosphate-buffered isotonic saline (PBS).
3. Cell suspension solution: 10 m*M* Tris-HCl, pH 7.2, 20 m*M* NaCl, and 100 m*M* ethylenediaminetetraacetic acid (EDTA).

4. InCert agarose (FMC Bioproducts): 1.6% in water. Dissolve using a microwave oven or boiling water bath and equilibrate to 65°C just prior to use. It may be stored refrigerated and remelted.
5. Proteinase K solution: 1 mg/mL proteinase K, 100 m*M* EDTA, pH 8.0, 1% sodium N-laurylsarcosine and 0.2% sodium deoxycholate.
6. Wash solution: 20 m*M* Tris-HCl, pH 8.0, and 50 m*M* EDTA. Supplement third wash with 1 m*M* phenylmethylsulfonyl fluoride (PMSF) using a freshly prepared solution of 100 m*M* PMSF in isopropanol.

2.4. RecA-Assisted Ligation

You need reagents listed in **Subheading 2.1.** plus:

1. Restriction endonucleases. Depending on the application, either one or two different enzymes may be used; and at least one must produce a 3'-recessed end.
2. DNA containing a fragment of interest.
3. A second DNA fragment to be ligated to the fragment of interest.
4. Exonuclease-free mutant of the Klenow fragment of *Escherichia coli* DNA polymerase I (New England Biolabs).
5. dATP, dCTP, dGTP, and TTP.
6. *E. coli* DNA ligase (*see* **Note 5**). If purchased from New England Biolabs, use the buffer provided with the ligase.
7. BSA: 10 mg/mL.
8. Chroma Spin + TE-400 gel filtration columns (from CLONTECH).

2.5. RecA-Assisted Cloning

You need reagents listed in **Subheadings 2.1.** and **2.4.**, except that:

1. The fragment of interest typically is from genomic DNA. To minimize the amount of DNA in the protocol (70 µg of starting DNA per clone obtained of a single-copy human genomic fragment and 3 µg if starting with yeast DNA), one should fractionate the DNA on a gel after cleavage with restriction endonucleases.
2. The second DNA fragment is a vector that has cohesive ends complimentary to the fragment of interest. Typically, the vector is also dephosphorylated.
3. T4 DNA ligase. If purchased from New England Biolabs, use the buffer provided with the ligase. 67 U of ligase from New England Biolabs is 1 Weiss U.
4. Transformable bacteria.
5. A method to remove unligated vector from the ligated fragment and vector is also needed, because the large amount of unligated vector will cause an unacceptably high background. One can imagine gel electrophoresis methods to perform this, but the only published method has used streptavidin beads (Dyna beads M–280, Dynal) coated with a short biotinylated duplex that contains a cohesive end complimentary to the selected fragment. When using the streptavidin bead technique, one also needs: a wash solution of 1 *M* NaCl in 50 m*M* Tris-HCl, pH 7.5, and glycogen (10 mg/mL). Immediately before use, the beads are saturated with the biotinylated duplex and excess duplex is removed by washing with the NaCl and Tris buffer.

3. Methods

With all RecA protein techniques, a synaptic complex is formed at 37°C. After formation, tubes may be withdrawn from the heating block to add the DNA-modifying enzymes, but care should be taken to minimize cooling, as the RecA protein can irreversibly dissociate at lower temperatures.

3.1. Protection of a Selected DNA Site from Restriction Endonuclease Cleavage: The Synaptic Complex Reaction

This protocol is for treatment of up to 1 µg of DNA, but may be scaled up to any volume desired if more than 1 µg is treated (*see* **Note 6**).

1. Add sufficient water to make the final volume 40 µL, 8 µL of RecA protein buffer (5X stock), 4 µL of ADP and ATPγS (10X stock), 0.18 µg of oligonucleotide (or 0.09 µg each of two oligonucleotides), 10 µg of RecA protein, and up to 1 µg of duplex DNA target. Add reagents in the above order and mix after each addition.
2. Incubate at 37°C for 10 min.
3. Add 20 U of restriction endonuclease, and incubate at 37°C for 20 min. This time may be increased to 1 hr if necessary for complete digestion.
4. Stop the reaction with 3 µL of 10% SDS and examine the products by gel electrophoresis.

3.2. RARE Cleavage

See **Note 7** before beginning.

1. Perform **steps 1** and **2** as in **Subheading 3.1.**
2. Add 2 µL of BSA, methylase (generally 4–8 U), and 2 µL of S-adenosylmethionine.
3. Incubate at 37°C for 20 min. This time may be increased to 1 h if necessary for complete methylation.
4. Incubate at 65°C for 15 min to inactivate the methylase and RecA protein.
5. Perform **steps 3** and **4** as above (*see* **Note 3**).

3.3. Preparation of DNA Embedded in Agarose

This procedure is used when working with large DNA fragments that tend to shear in solution. Fragments smaller than 50 kb do not need to be embedded, and shearing may be minimized by using wide-bore pipet tips and gentle technique.

1. Wash cells (1 × 10^8, about 150 mg wet weight) twice with PBS and suspend in 1 mL of suspension buffer.
2. Incubate for 1 min at 65°C.
3. Mix in 1 mL of InCert agarose at 65°C.
4. Transfer the mixture to plug molds or the barrels of 1-mL syringes and allow to set.
5. Transfer the gel to a test tube and incubate in 6 mL of proteinase K solution for 20 h at 42°C with gentle shaking.

6. Remove solution, wash twice with 10 mL of wash solution for 15 min at room temperature.
7. Wash once for 1 h with wash solution supplemented with PMSF.
8. Wash twice with wash solution alone for 15 min each and twice with a tenfold dilution of wash solution.
9. Store DNA at 4°C.

3.4. RARE Cleavage of DNA Embedded in Agarose

1. Using a cover slip, finely chop agarose containing DNA. Use 25–50 µL agarose per reaction.
2. Soak chopped agarose in 50 m*M* EDTA, pH 8.0 for 30 min.
3. Remove EDTA and soak chopped agarose in 1X RecA protein buffer for 5 min. Exchange RecA protein buffer four additional times to remove all EDTA. Suspend agarose in about 100 µL total volume per reaction.
4. Add sufficient water to make the final volume 200 µL (including the step 6 components), 20 µL RecA protein buffer (5X stock), 4 µL EGTA, 20 µL ADP and ATPγS (10X stock), 0.36 to 1.2 µg oligonucleotide (0.18 to 0.6 µg each of two oligonucleotides), 30 µg RecA protein, and 100 µL of the agarose suspension. Routinely, three or four individual reactions covering the above range of oligonucleotide concentrations are performed. Add reagents in the above order and mix after each addition.
5. Incubate at 37°C for 15 min.
6. Add 10 µL BSA, methylase (generally 80 U), and 10 µL S-adenosylmethionine.
7. Incubate at 37°C for 1–2 h.
8. Stop reaction with 20 µL SDS and incubate for 15 min at 37°C.
9. Exchange into any desired restriction endonuclease buffer as in **step 3**. Leave about 150 µL buffer after the last exchange.
10. Add 80 U of a restriction endonuclease and incubate for 1 h.
11. If desired, the reaction may be stopped with SDS and examined by gel electrophoresis. The chopped agarose may be loaded in the well of a gel using a wide-bore pipet tip.

3.5. RecA-Assisted Ligation

This protocol is for treatment of up to 2.5 µg of DNA, but may be scaled up to any volume desired if more than 1 µg is treated (*see* **Note 6**).

1. Add sufficient water to make the final volume 100 µL, 20 µL RecA protein buffer (5X stock), 10 µL ADP and ATPγS (10X stock), 0.32 µg oligonucleotide (or 0.16 µg each of two oligonucleotides), 13 µg RecA protein, and up to 2.5 µg of duplex DNA target. Add reagents in the above order and mix after each addition.
2. Incubate for 10 min at 37°C.
3. Add 4 µL BSA (10 mg/mL stock), 12.5 U of the polymerase, and each of the 4 deoxynucleoside triphosphates to a final concentration of 38 µ*M* each.
4. Incubate for 30 min at 37°C.

5. Stop the reaction with 10 µL SDS or, if proceeding to the next step, by phenol extraction and ethanol precipitation. Heating to 65°C will not adequately stop the polymerase.
6. A short duplex may be ligated to the selected fragment (*see* **Note 5**) by dissolving the duplex DNA target in 100 µL *E. coli* ligase buffer containing 2 mg of a short duplex and 20 U *E. coli* ligase for 1 h at room temperature.
7. If desired, remove unligated short duplex by gel filtration.

3.6. RecA-Assisted Cloning

This protocol is for cloning directly from human DNA into a plasmid vector (*see* **Note 8**), but may be scaled down by a factor of 20 when cloning genomic fragments from yeast or bacteria.

1. Add sufficient water to make the final volume 1.44 mL, 0.29 mL of RecA protein buffer (5X stock), 0.144 mL ADP and ATPγS (10X stock), 3.2 mg each of the two oligonucleotides, 360 µg RecA protein, and about 26 µg DNA target. Add reagents in the above order and mix after each addition (*see* **Note 9**).
2. Incubate for 10 min at 37°C.
3. Add 57 µL BSA (10 mg/mL stock), 450 U of the polymerase, and each of the 4 deoxynucleoside triphosphates to a final concentration of 38 µ*M* each.
4. Incubate for 30 min at 37°C.
5. Stop the reaction by phenol extraction and ethanol precipitation. Heating to 65°C will not adequately stop the polymerase.
6. Add 2 mg of coated beads, 80 µL T4 DNA ligase buffer, 2.6 µg of the processed human DNA (one-tenth of the material processed above), 3.4 µg vector, and 3200 U T4 DNA ligase and incubate for 16 h at 16°C to form the complex shown in **Fig. 3**.
7. Wash beads three times.
8. Add 0.3 µg of the lower strand of the biotinylated duplex to replace any lost by washing.
9. Exchange the beads into any desired buffer and cleave the fragment vector from the short biotinylated duplex using the appropriate restriction endonuclease (*Eco*RI in the example shown in **Fig. 3**).
10. Add 2 µL of glycogen.
11. Perform ethanol precipitation.
12. Circularize the construct using 1600 U T4 ligase in 100 µL of its buffer for 16 h at 16°C.
13. Concentrate the material by ethanol precipitation and use to transform a suitable bacterial host.
14. Screen colonies for the correct clone or use pooled colonies as the starting DNA for a repeat round of RecA-assisted cloning.

4. Notes

1. When making this buffer, one should remember that the correct pH is important for RecA function and that the pH of Tris buffer stocks change upon dilution. 0.50 *M* Tris acetate at a pH of 8.00 is a convenient stock to use.

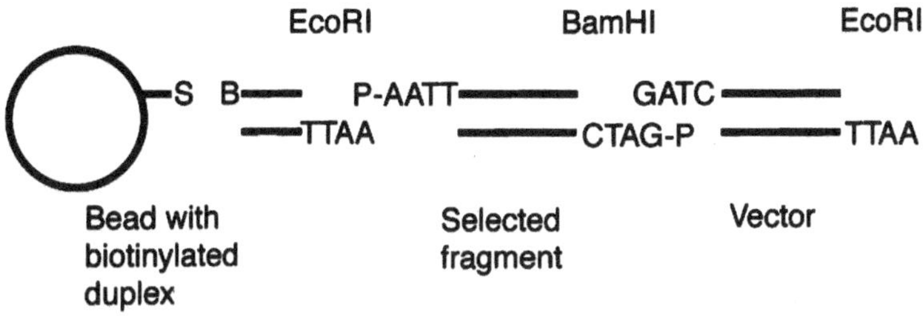

Fig. 3. Construct made when selected fragment-vector molecules are separated from unligated vector. *Eco*RI and *Bam*HI cohesive ends are shown. **S**, streptavidin; **B**, biotin; **P**, 5' phosphate.

2. The oligonucleotides used to form synaptic complexes should have the same sequence and polarity as either one of the duplex target strands, although the yield is slightly higher if the strand highest in purines is used. They should be 30 –40 bases long, although a slightly higher yield is obtained with 60 bases. Oligonucleotides shorter than 40 bases are routinely used without purification, but any residual solutes remaining after synthesis are removed by ethanol precipitation. When blocking restriction endonucleases or methylases, the recognition sequence of the enzyme should be near the middle of the oligonucleotide. If the sequence on only one side of the recognition sequence is known, choose the strand such that the recognition sequence is at the 5'-end of the oligonucleotide and extend that end with about 10 bases of any sequence. In RecA-assisted ligation or RecA-assisted cloning, there is no need to use oligonucleotides longer than 30 bases, the sequence at the end of either strand may be used, and there is no need to add an extension. Determine the concentration of each oligonucleotide assuming that 1 OD_{260} unit is 33 µg.
3. When doing a synaptic complex protection, the restriction endonuclease used must have at least some activity in the RecA protein buffer. Generally, any endonuclease with some activity at low-salt concentration will function. An alternative RecA protein buffer consisting of 100 m*M* Tris-HCl, pH 7.5, 62.5 m*M* $MgCl_2$, 2 m*M* DTT (5X stock) may be used if necessary to increase activity. When doing RARE cleavage, if the endonuclease used is not active in the RecA protein buffer, one may dissolve the methylated DNA in any buffer of choice. When using *Eco*RI to digest large fragments of DNA, a buffer of 100 m*M* Tris-HCl, pH 8.0, 50 m*M* NaCl, 20 m*M* $MgCl_2$, 1.5 µ*M* DTT, and 200 µg/mL BSA is useful for minimizing star activity.
4. New England Biolabs carries the largest selection of methylases. *Eco*RI methylase works well for the most demanding applications involving large fragments of genomic DNA. *Alu*I, *Bam*HI, *Taq*I, *Sss*I, *Hpa*II, and *Hha*I methylases can be used to create fragments at least 50 kb in size. *Hae*III and *Dam* methylases are not very active in the RecA protein buffer, but could be used when making short fragments requiring methylation of only a couple of intervening sites.

5. *E. coli* DNA ligase has a high specificity for joining DNA with cohesive ends, but commercial preparations are not very concentrated. The activity is sufficient when one of the DNA fragments is present at a relatively high concentration, such as when a short synthetic duplex is used. When joining two fragments that are both present at low concentration, the more concentrated but less specific T4 DNA ligase should be used.
6. All RecA protein techniques may be scaled up, and total volumes as high as 1.5 mL have been used. The concentration of the duplex DNA target may be decreased if desired, but the concentrations of all the other components should be maintained.
7. Given the large number of reagents used in RARE cleavage, it is useful to perform two preliminary experiments to test the reagents. The first is to perform a synaptic complex protection as in **Subheading 3.1**. This will test a large number of the reagents such as the RecA protein and the oligonucleotide. The second is to leave out the RecA protein and oligonucleotide and test a range of methylase concentrations. Especially with expensive methylases, one should minimize the amount of methylase added. Any incomplete methylation should be evident after treatment with the appropriate restriction endonuclease. Methylases typically are supplied in 50% glycerol, and as in all RecA protein techniques, the final glycerol concentration should not exceed 5%.
8. While the example described in the literature and presented here describes a selected *Eco*RI-*Bam*HI fragment cloned into a plasmid vector, the only requirements of the technique are that at least one end of the fragment has a 3'-recessed end and that the vector (be it a plasmid, BAC, or PAC vector) has at least one complimentary cohesive end.
9. The only example described so far in the literature *(1)* used the technique to clone a human genomic DNA fragment. The DNA had been digested with *Eco*RI and *Bam*HI and fractionated on an agarose gel. Overall, the technique yielded one correct clone for each 70 µg of starting DNA, and the enrichment of the technique was 1600 fold. In this case, a plasmid was used as the second DNA substrate, and highly efficient electrocompetent bacterial cells were used as the cloning host.

References

1. Ferrin, L. J. and Camerini-Otero, R. D. (1998) Sequence-specific ligation of DNA using RecA protein. *Proc. Natl. Acad. Sci. USA* **95,** 2152–2157.
2. Ferrin, L. J. and Camerini-Otero, R. D. (1994) Long-range mapping of gaps and telomeres with RecA-assisted restriction endonuclease (RARE) cleavage. *Nat. Genet.* **6,** 379–383.
3. National Institutes of Health and Institute of Molecular Medicine Collaboration. (1996) A complete set of human telomeric probes and their clinical application. *Nat. Genet.* **14,** 86–89.
4. Reston, J. T., Hu, X., Macina, R. A., et al. (1995) Structure of the terminal 300 kb of DNA from human chromosome 21q. *Genomics* **26**, 31–38.

5. Feder, J. N., et al. (1996) A novel MHC class 1-like gene is mutated in patients with hereditary haemochromatosis. *Nat. Genet.* **13,** 399–408.
6. Lauer, P., Schneider, S. S., and Gnirke, A. (1998) Construction and validation of yeast artificial chromosome contig maps by RecA-assisted restriction endonuclease cleavage. *Proc. Natl. Acad. Sci. USA* **95,** 11,318–11,323.
7. Boren, J., Lee, I., Callow, M. J., et al. (1996) A simple and efficient method for making site-directed mutants, deletions, and fusions of large DNA such as P1 and BAC clones. *Genome Res.* **6,** 1123–1130.
8. Nielsen, L. B., Kahn, D., Duell, T., et al. (1998) Apolipoprotein B gene expression in a series of human apolipoprotein B transgenic mice generated with recA-assisted restriction endonuclease cleavage-modified bacterial artificial chromosomes. *J. Biol. Chem.* **273,** 21,800–21,807.
9. McClelland, M. and Nelson, M. (1992) Effect of site-specific methylation on DNA modification methyltransferases and restriction endonucleases. *Nucleic Acids Res.* **20S,** 2145–2157.
10. Smith, C. L., Klco, S., Zhang, T., et al. (1993) Analysis of megabase DNA using pulsed-field gel electrophoresis techniques. *Methods in Molecular Genetics, Volume 2B* (Adolph, K. W., ed.), Academic, San Diego, CA, pp. 155–194.
11. McClelland, M. (1987) Site-specific cleavage of DNA at 8-,9-, and 10-bp sequences. *Methods in Enzymology, Volume 155* (Wu, R., ed.), Academic, San Diego, CA, pp. 22–32.
12. Carle, G. F. and Olson, M. V. (1987) Orthogonal-field-alternation gel electrophoresis, in *Methods in Enzymology, Volume 155* (Wu, R., ed.), Academic, San Diego, CA, pp. 468–482.

12

Interspecies Recombination and Mismatch Repair

Generation of Mosaic Genes and Genomes

Ivan Matic, François Taddei, and Miroslav Radman

1. Introduction

The efficiency of homologous genetic recombination depends on the length and degree of DNA sequence identity shared by the two parental molecules, and on the cellular enzymatic systems involved in genetic recombination and its editing. Crosses between different bacterial species show that recombination frequency decreases exponentially with increasing sequence divergence *(1)*. Recombination between DNAs differing by large blocks of heterology is quite efficient, whereas recombination between DNAs differing by well-spaced point mutations is reduced or inhibited *(2)*. These observations suggest that the main determinant of genetic isolation between genomes of closely related bacterial species is neutral sequence divergence caused by point mutations, rather than large heterologies.

The extent and degree of homology determine the activities of the enzymes that control the fidelity of recombination *(2)*. The initiation of recombination in bacteria depends on the activity of the RecA protein, the level of which is controled by the SOS system *(3)*. The SOS system is a set of physiological responses that is induced by exposure of bacterial cells to a variety of conditions that damage DNA and/or interfere with its replication (ref). This inducible genetic system has two features that favor the survival of stressed bacterial populations: it enhances the capacity for DNA repair and it increases genetic variation *(3a)*. The induction of the SOS system increases RecA concentration, as well as the frequency of recombination (*see* **Note 1**). RecA is highly selective for sequence identity only at the initial stage of the strand exchange pro-

From: *Methods in Molecular Biology, vol. 152: DNA Repair Protocols: Prokaryotic Systems*
Edited by: P. Vaughan © Humana Press Inc., Totowa, NJ

cess, which requires a minimum length of homology or Minimum Efficient Processing Segment (MEPS), below which recombination becomes inefficient *(4)*. In *Escherichia coli*, MEPS is about 23–27 bp. Divergence between recombination substrates reduces recombination by reducing the number of available MEPSs. However, once initiated, RecA-mediated strand exchange can occur in spite of large numbers of mismatches and large heterologies, as demonstrated in vitro *(5,6)*. During this stage of genetic recombination, the fidelity is controled by the methyl-directed mismatch repair (MMR) system. Mismatch recognizing and binding protein, MutS, recognizes mispaired and unpaired bases in the joint heteroduplex regions and blocks RecA-catalyzed strand transfer *(7)*. Even a very low divergence is sufficient to impede recombination. The increased concentration of RecA protein does not prevent MMR editing activity *(3)*. The activity of RuvAB proteins, which can also catalyze branch migration across mismatched regions, appears not to be subject to such editing. The RuvAB-catalyzed heteroduplexes are either too long to be edited or RuvABC protein-mediated resolution is too fast for MutS and MutL to act.

The concentration of MutS and MutL proteins alters the apparent length of the segment of sequence identity needed for successful recombination *(1)*. Thus, MMR is a potent inhibitor of recombination between genomes of related bacterial species and between non-identical DNA sequences in general *(3)*. The inactivation of *mutS* or *mutL* genes results in 10^3-fold increase in conjugational and 10^2-fold increase in transductional recombination between *E. coli* and *Salmonella typhimurium,* whose chromosomes are about 16% divergent at the DNA sequence level *(8)*. Transductional recombination between *S. typhimurium* and *Salmonella typhi*, whose genomes differ only 1–2% at DNA sequence level, increase 10^2–10^3-fold in MMR-deficient genetic background *(9)*. Conversely, overproduction of MutS and MutL proteins severely reduces recombination frequency even between bacterial strains with very low genomic divergence *(1)*.

The current understanding of the molecular mechanisms controlling genetic exchange among bacteria involves two key enzymatic components: the MMR and SOS system *(1,8,10)*. By modifying their activities, it is possible to disrupt or establish genetic barriers between different bacterial species. One of the major problems with recombination in MMR proficient background is that DNA sequence divergence restricts recombination events to the genomic regions of highest sequence similarity. Consequently, the majority of recombinants are highly unstable merodiploids, often generated by an unequal crossover between *rrn* operons, which are usually almost identical between related bacterial species. The inactivation of MMR allows, not only higher yield of interspecies recombinants, but also higher diversity and higher stability of hybrids *(11)* (*see* **Note 2**).

The potential applications of this strategy, that allows creation of interspecific mosaic genomes, are multiple in fundamental research, biotechnology, genetic engineering, and medicine. By allowing recombination of entire genomes, the mosaic genomes with new combinations of genes and operons, or even with mosaic genes and operons can be obtained (*see* **Notes 3–6**). For example, this approach can facilitate modifications and improvement of very complex metabolic networks. Another possible application is creation of attenuated live vaccines by recombining pathogenic bacteria and their nonpathogenic counterparts, in order to create nonpathogenic hybrids that retain the immunogenicity of the parental strains.

In this chapter, the strategies that allow maximal yield of interstrain and interspecies recombinants are described. The conjugation provides the best mean of gene transfer because of large host range of conjugative plasmids, as well as the relative immunity of effective conjugational DNA transfer to host modification-restriction systems and host nucleases (*see* **Note 3**). The manipulation of the molecular mechanisms that controls recombination between divergent DNAs, particularly the two key components: MMR and SOS systems are described.

2. Materials

2.1. Bacterial Stains

2.1.1. Donor Strains

The most commonly used conjugative plasmid is F, which can replicate and express its conjugation system in *E. coli* and many other bacterial species. The sets of *E. coli* Hfr and F' derivatives, as well Hfr and F' strains with a Tn*10* insertions that provide easily selectable donor markers, are available from *E. coli* Genetic Stock Center, Yale University, New Haven, CT (http://cgsc.biology.yale.edu). Those strains can transfer any *E. coli* chromosomal loci (*see* **Notes 3** and **7**). However, the F plasmid cannot be used for all bacterial species. In such cases, conjugative plasmids, belonging to IncP alpha group, like RP4 can be used. These plasmids are remarkable for their broad host range and can mobilize chromosomal markers from any point of origin.

2.1.2. Recipient Strain Genotype

The alleles and their phenotypes described are *E. coli* derivatives. For the majority of other bacterial species, similar genetic backgrounds are not available. Therefore, if not otherwise possible (*see* **Notes 1–4**), *E. coli* strains with appropriate genetic background should be used as recipients. The highest frequency of interspecies recombination can be obtained using *E. coli* recipient strain with following alleles ***(1,3,8)***.

1. *mutS*-: MutS protein is the mismatch recognizing and binding component of *E. coli* MMR. Inactivation of this gene increases, up to 10^3-fold, the frequency of recombination between diverged DNA sequences *(8)* (*see* **Note 2**).
2. *recAoc*: This allele constitutively increases RecA protein concentration, thus stimulating conjugational intra- and interspecies recombination *(3)*. The mutation in the operator of *recA* gene prevents its repression by the LexA repressor. An increase in RecA concentration appears to increase the total length of DNA available for recombination *(1)* (*see* **Note 1**).
3. *recD*-: The inactivation of this gene results in small increase of conjugational recombination *(3)*. The effect is not as strong as during transduction because the size of transferred DNA and the high number of Chi sequences attenuate RecBCD exonuclease, thus rendering donor DNA relatively immune to degradation by the RecBCD exonuclease (*see* **Note 3**).
4. *hsdR*-: This gene encodes *E. coli* type I restriction endonuclease, whose inactivation results in a small increase of conjugational recombination *(3)* (*see* **Note 4**).

2.2. Media

1. LB rich medium: 10 g/L bacto tryptone (Difco), 5 g/L bacto yeast extract (Difco), 10 g/L NaCl, deionized H_2O to 1 L. Adjust pH to 7.0 with 1 *M* NaOH.
2. M63 minimal medium: 13.6 g/L KH_2PO_4, 2 g/L $(NH_4)_2SO_4$, 0.5 mg/L $FeSO_4$ $7H_2O$, deionized H_2O 1 L. Adjust pH to 7.0 with KOH. After autoclaving, add 1 mL of 1 *M* $MgSO_4{\cdot}7\ H_2O$, and 10 mL of 20% solution of carbon source per liter. Vitamins, such as B1, are added to a final concentration of 1 μg/mL and amino acids at 80 μg/mL.

If media are used in plates, add 15 g agar (Difco) per liter. All media should be sterilized by autoclaving for 30 min at 120°C. Antibiotics are always added after autoclaving. Stock solution of nalidixic acid is 100 mg/mL in 1 *N* NaOH. Stock solution of streptomycin is 100 mg/mL in H_2O. Stocks of antibiotics should be sterilized by filtration and kept at –20°C.

3. Methods

3.1. Growth of Cells Prior to Conjugation

The condition of donor cells in conjugation is much more crucial than that of recipient strain. In order to obtain the donor cells in physiological state optimal for conjugation *(12,13)*.

1. Dilute the overnight culture 50-fold into fresh LB medium and grow with gentle shaking (about 150 rpm). Pregrowth in rich medium, such as LB, results in much higher efficiencies of conjugation than growth in minimal medium. Pregrowth should be at 37°C, because donors that are pregrown, for example, at 33°C, produce very few F pili and the mating efficiency is only 10% of those at 37°C.

2. The cultures of donor and recipient strains should reach OD_{600} 0.3–0.4 (2–3 × 10^8 cells/mL). Cultures that are at a higher density mate less efficiently. At this stage, donor and recipient cells can be kept on ice for several hours before starting the cross.

3.2. Conjugational Crosses

The conjugation on solid support is much more efficient than conjugation in liquid medium.

1. Mating mixtures are prepared by mixing together Hfr (donor) and F^- (recipient) bacteria in a 1:1 ratio. Usually, 1 mL of donor cells is mixed with 1 mL of recipient cells, however, this can be scaled up if required. This mixture is filtered through a sterile 0.45-μ*M* pore-size nitrocellulose filter (2.5-cm diameter), using vacuum filtration equipment (Millipore). It is very difficult to put more than about 5 × 10^8 cells per filter because such cellular mass prevents further filtration.
2. The filters with mating mixtures are incubated on fresh LB agar plate, which are prewarmed at 37°C. Conjugation efficiency is highest at this temperature and falls off sharply at temperatures below 30°C or above 40°C.
3. After incubation for the required period of time, mating pairs are resuspended in 10^{-2} *M* $MgSO_4$ and separated by vigorous vortexing. If matings must be stopped quickly, nalidixic acid or streptomycin can be added to 10^{-2} *M* $MgSO_4$ at concentrations of 50 μg/mL and 100 μg/mL, respectively. At this stage, donor and recipient cells can be kept on ice for several hours before plating. The incubation time used in this step is dependent on the marker being transferred. The speed of transfer is approx 47 kb/min. The whole *E. coli* chromosome (approx 4.7 Mb) is expected to be transferred in 100 min. Therefore, if a given marker is 470 kb from the origin of transfer, then mating is carried out for 15 min in order to ensure that this particular marker is transferred. The position of the origin of transfer varies with different Hfr stains, therefore the time of conjugation of a given marker will vary from strain to strain.

3.3. The Selection of Recombinants

The mating mixtures are subsequently plated on medium that selects for the desired recombinants and counterselects both donor and recipient parental strains. Interspecies recombinants often grow slower than parental strains and intraspecies recombinants. When minimal medium is used for selection, up to 4 d of incubation is necessary to obtain maximal yield of recombinants. The markers that can be used for selection and/or counterselection after conjugation are sensitivity or resistance to antibiotics, auxotrophic growth requirements, sensitivity or resistance to bacteriophage infection, and temperature- or cold-sensitive growth (*see* **Notes 8** and **9**).

The best recipient markers used for counterselection against donor cells are nalidixic acid or streptomycin resistance. The level of spontaneous resistance to these antibiotics is very low, about 5 × 10^{-10} for streptomycin and about 10^{-9}

for nalidixic acid. Nalidixic acid stops matings quickly and prevents remating on plates. Streptomycin does not stop the mating quickly and matings can continue on selective plates. Those two antibiotics are used at concentrations of 100 µg/mL or 50 µg/mL, respectively. Both antibiotics can be used in rich and minimal media. It is necessary to avoid the transfer of chromosomal regions carrying genes whose mutations render cells resistant to the antibiotic used to counterselect donor cells. Resistance to streptomycin is encoded by the modified *rpsL* gene and nalidixic acid by the modified *gyrA* gene, at 78.4 min and 50.3 min of *E. coli* genetic map, respectively.

4. Notes

1. The induction of the SOS system was detected during conjugation between different enterobacterial species, and the level of induction increases with the increase of DNA sequence divergence [Matic, I., unpublished data and *(3)*]. Genetic characterization of the SOS induction during *E. coli* × *S. typhimurium* matings suggested that the strong and persistent activation of RecA is probably caused by the reduced efficiency of early RecA-catalyzed recombination steps as a consequence of DNA divergence. The SOS induction is necessary for efficient intra- and interspecies recombination. Therefore, if DNA divergence is low, the recipient strains can be UV irradiated in order to induce SOS system prior to conjugation, thus increasing concentration of RecA, RuvA, and RuvB proteins and alleviating host-restriction systems. Concentration of RecA protein can also be increased by using plasmid carrying *recA* gene under control of arabinose or tac-inducible promoters. Using such plasmids, concentration of RecA protein can vary, in the *recA* background, from 0 to 10,000 molecules per cell. This might be particularly important for stabilization of interspecies hybrids by preventing further RecA-dependent rearrangements.
2. The efficiency of the MMR can be decreased as a result of titration of the mismatch repair proteins. MMR is saturated in cells that are treated with DNA damaging agents introducing DNA lesions recognized by MutS protein. This is probably why the frequency of interspecies transductional recombination can be increased by treating recipient cells with MNNG *(17)*. MutS protein can also be titrated in cells containing mismatches in retron-encoded multicopy single-stranded DNA *(18)*. Overexpression of such molecules results in increased mutagenesis and frequency of interspecies recombination. The advantage of transient saturation of MMR is that, after recombination, hybrid cells do not have constitutively high mutation rates because of inactivation of MMR genes. This strategy can be particularly appropriate for bacterial species whose mismatch repair deficient mutants is difficult to construct.
3. The choice of DNA transfer method depends on the length of DNA that should be involved in recombination, as well as on bacterial DNA species that serve as a recipient. Conjugative plasmids mobilise megabases of donor DNA. Transducing phages transfer tens to hundreds of kilobases. Segments of transforming DNA are

several kilobases long. When transduction is used as the mean of genetic exchange, inactivation of *recD* gene is necessary for efficient interspecies recombination ***(14)***. By degrading donor DNA, RecBCD exonuclease reduces the frequency of transductional recombination between *S. typhimurium* and *S. typhi* by 10^1–10^2-fold compared to the frequency observed in *recD-* genetic background ***(14)***.

4. The restriction of donor DNA can stimulate recombination, because DNA ends generated by restriction enzymes are recombinogenic. The restriction systems in *E. coli* can influence the pattern of DNA fragments incorporated into the recipient chromosome, creating mosaics of short sequences. However, if the size of transferred donor DNA is small or if restriction fragments are small, they can be completely degraded by exonucleases. Hence, inactivation of *hsdR* gene in recipient strain strongly stimulates transductional recombination. The activity of most of restriction endonucleases in *E. coli,* and other bacterial species can be alleviated upon exposure of cells to UV light, heat, organic solvents, pH shifts, or detergents ***(15,16)***. Restriction enzymes can be titrated by large amounts of donor DNA introduced during prolonged conjugation.
5. The strategy for recombining diverged genomes described in this chapter can be applied as well to recombined cloned diverged genes. For example, when 405 bp DNA segments with 89% of sequence identity, cloned in lambda phage and pBR322 plasmid, were recombined, the frequency of recombination was 15-fold higher in *mutS* deficient that in the wild-type background ***(19)***. Using this approach, it would be possible to combine functional domains from different existing enzymes and incorporate them in the same molecule in order to generate the proteins and enzymes with novel functions and properties. This can be particularly successful with enzymes that have modular structure, as it has been already demonstrated by in vitro manipulation of polyketide-synthase enzymes ***(20)***.
6. Recombining genes cloned in plasmid or phage vectors without flanking homology has the advantage that recombination takes place entirely within defined DNA sequences. The genes should be cloned in inverted orientation in order to allow multiple rounds of reciprocal crossover. However, at a high-sequence divergence between recombining genes, no recombination initiation site (MEPS) would be present. Therefore, such highly diverged genes would be cloned along with 50–100 bp of sequence identity to provide for initiation of RecA-catalyzed strand exchange which, under MMR deficiency conditions, would extend into the diverged region.
7. Many Hfr strains are unstable for Hfr character and accumulate a substantial proportion of F^+ revertants. Similarly, F' factors are frequently lost from their host upon storage. Therefore, it is necessary to occasionally check the stability of donor strains and repurify the old stocks ***(13)***.
8. The choice of selective and counterselective markers can influence the background growth resulting from a bacterial cross. Both leakiness and reversion contribute to the background growth. In general, vitamin markers are much leakier than amino acid or sugar markers. Crossfeeding and inactivation of antibiotic or other inhibitors can also affect background growth.

9. Antibiotics can be used as markers for selection after conjugation, but many antibiotic resistance alleles are recessive and several generations of growth would be required to allow segregation of, for example, rifampicin-resistant recombinants. Because it is more convenient to plate mating mixtures out directly, it is preferable to use those markers that allow selection following direct plating.

References

1. Vulic, M., Dionisio, F., Taddei, F., and Radman, M. (1997) Molecular keys to speciation: DNA polymorphism and the control of genetic exchange in enterobacteria. *Proc. Natl. Acad. Sci. USA* **94,** 9763–9767.
2. Radman, M. and Wagner R. (1993) Mismatch recognition in chromosomal interactions and speciation. *Chromosoma* **102,** 369–373.
3. Matic, I., Rayssiguier C., and Radman M. (1995) Interspecies gene exchange in bacteria: the role of SOS and mismatch repair systems in evolution of species. *Cell* **80,** 507–515.

3a. Friedberg, E. C., Walker, G. C., and Siede, A. (1995) *DNA Repair and Mutagenesis*, ASM Press, Washington, DC.

4. Shen, P. and Huang, H. V. (1986) Homologous recombination in *Escherichia coli*: dependence on substrate length and homology. *Genetics* **112,** 441–457.
5. DasGupta, C. and Radding, C. M. (1982) Polar branch migration promoted by RecA protein: effect of mismatched base pairs. *Proc. Natl. Acad. Sci. USA* **79,** 762–766.
6. Bianchi, M. E. and Radding, C. M. (1983) Insertions, deletions, and mismatches in heteroduplex DNA made by RecA protein. *Cell* **35,** 511–520.
7. Worth, Jr., L., Clark, S., Radman, M., and Modrich, P. (1994) Mismatch repair proteins MutS and MutL inhibit RecA-catalyzed strand transfer between diverged DNAs. *Proc. Natl. Acad. Sci. USA* **91,** 3238–3241.
8. Rayssiguier, C., Thaler, D. S., and Radman, M. (1989) The barrier to recombination between *Escherichia coli* and *Salmonella typhimurium* is disrupted in mismatch-repair mutants. *Nature* **342,** 396–401.
9. Zahrt, T. C., Mora, G. C., and Maloy, S. (1994) Inactivation of mismatch repair overcomes the barrier to transduction between *Salmonella typhimurium* and *Salmonella typhi*. *J. Bacteriol.* **176,** 1527–1529.
10. Matic, I., Taddei, F., and Radman, M. (1996) Genetic barriers among bacteria. *Trends Microbiol.* **4,** 69–73.
11. Matic, I., Radman, M., and Rayssiguier, C. (1994) Structure of recombinants from conjugational crosses between *Escherichia coli* donor and mismatch-repair deficient *Salmonella typhimurium* recipients. *Genetics* **136,** 17–26.
12. Miller, J. H. (1992) A Short Course in Bacterial Genetics. *A Laboratory Manual for* Escherichia coli *and Related Bacteria*, Cold Spring Harbor Laboratory Press, Cold Spring Harbor, NY.
13. Low, K. B. (1991) Conjugational methods for mapping with Hfr and F-prime strains, in *Bacterial Genetic Systems* (Miller, J. H., ed.), Academic, San Diego, CA, pp. 43–62.

14. Zahrt, T. C. and Maloy, S. (1997) Barriers to recombination between closely related bacteria: MutS and RecBCD inhibit recombination between *Salmonella typhimurium* and *Salmonella typhi. Proc. Natl. Acad. Sci. USA* **94,** 9786–9791.
15. Schafer, A., Kalinowski, J., and Puhler, A. (1994) Increased fertility of Corynebacterium glutamicum recipients in intergeneric matings with *Escherichia coli* after stress exposure. *Appl. Environ. Microbiol.* **60,** 756–759.
16. Kelleher, J. E. and Raleigh, E. A. (1994) Response to UV damage by four *Escherichia coli* K-2 restriction systems. *J. Bacteriol.* **176,** 5888–5896.
17. Mergeay, M. and Gerits, J. (1983) Transduction of *Escherichia coli* trp genes in *Salmonella typhimurium* and effect of N-methyl-N'-nitro-N-nitrosoguanidine on transduction with heterogenotic DNA. *J. Gen. Microbiol.* **129,** 321–335.
18. Maas, W. K., Wang, C., Lima, T., et al. (1996) Multicopy single-stranded DNA of *Escherichia coli* enhances mutation and recombination frequencies by titrating MutS protein. *Mol. Microbiol.* **19,** 505–509.
19. Shen, P. and Huang, H. (1989) Effect of base pair mismatches on recombination via the RecBCD pathway. *Mol. Gen. Genet.* **218,** 358–360.
20. Oliynyk, M., Brown, M. J., Cortes, J., et al. (1996) A hybrid modular polyketide synthase obtained by domain swapping. *Chem. Biol.* **3,** 833–839.

13

The Use of Immobilized Mismatch Binding Protein in Mutation/SNP Detection

Robert Wagner and Alan Dean

1. Introduction

The detection of single-base change mutations and polymorphisms is of enormous importance, both in research and in diagnostics. The ability to identify and score single nucleotide polymorphisms (SNPs) is becoming a key element of gene identification and mapping, and the future of human diagnostics depends on having the ability to detect single-base change mutations, because these represent the vast majority of disease-causing and disease-associated mutations. An ideal system for detecting and scoring these mutations and SNPs will have certain key characteristics: (1) robustness: the method will be user friendly and not subject to wide fluctuations caused by small changes in experimental conditions; (2) high throughput: given the requirements of genomic research and large-scale diagnostics, a useful method of mutation/SNP detection must be able to handle thousands of samples per day with limited technician effort; (3) low cost: for wide-spread use in both research and clinical diagnostics, low-cost and easy availability of both equipment and reagents is crucial; (4) no gels: this requirement is primarily to meet the high throughput requirement; (5) no radioactivity: given the problems of radioactive material handling and disposal and the availability of a wide variety of alternatives, radioactivity should not be a part of the ideal mutation detection system. Although there may be additional preferences of individual researchers, any mutation/SNP detection system that successfully meets all of the above requirements will enjoy wide-spread application. Immobilized mismatch binding protein (IMBP) is such a detection system.

From: *Methods in Molecular Biology, vol. 152: DNA Repair Protocols: Prokaryotic Systems*
Edited by: P. Vaughan © Humana Press Inc., Totowa, NJ

The IMBP mutation detection technology is based on MutS, the mismatch recognition and binding component of the *Escherichia coli* mismatch repair system [for review, *see* ***(1)***]. Homologs of MutS have been found in all species that have been examined, including bacteria, yeast, frogs, mice, and humans. *E. coli* MutS is a 97-kDa protein and appears to function as a dimer in mismatch recognition.

The *E. coli* mismatch repair system does not recognize and repair all mismatches with equal efficiency ***(2)***. [A similar recognition pattern is displayed by MutS in vitro ***(5)***]. Somewhat counterintuitively, the system recognizes and repairs most readily those mismatches that most resemble basepairs, i.e., those that cause the least helical distortion. It has been demonstrated in NMR studies that repairable mismatches are able to adopt intrahelical configurations and generally exhibit a fair degree of stacking and basepairing ***(4)*** . In addition, mismatches in G:C rich regions are more readily repaired than mismatches in A:T rich regions ***(5)***. This mismatch-recognition pattern is perfectly complementary to the pattern of polymerase error generation, i.e., polymerase errors occur most frequently in G:C rich regions and resemble basepairs. In addition, polymerase tends to make small additions or deletions, particularly in regions of repeated nucleotides. Mismatch repair recognizes and repairs frameshift/wild-type heteroduplexes containing up to four unpaired bases ***(6)*** (it should be noted that up to four unpaired bases can exist in an intrahelical configuration). In fact, a single unpaired base heteroduplex is repaired as well as the best repaired mismatch. The mismatch recognition pattern of MutS makes it ideally suited to a mutation detection system and distinguishes it from other mutation detection systems that depend on detection of mismatches. Methods that require mismatch cleavage depend on detecting distortion in the DNA helix, as do methods that depend on oligonucleotide annealing, such that mismatches that most resemble basepairs (e.g., transition mutations in G:C rich regions) and that are, therefore, among the most commonly found, are the most difficult to detect.

IMBP assays have been successful in a variety of formats, including the original nitrocellulose format ***(7)***, magnetic beads ***(8)***, and 96-well plates ***(9)***. All formats of IMBP perform significantly better than MutS in solution and can detect all single-base change mutations, as well as insertion/deletion mutations of up to four basepairs. The most user-friendly and cost-effective format is the microtiter plate format. At present, only 96-well plates are available, but, with automation, 384-well plates, and perhaps even 1536-well plates, should be possible. This chapter focuses on the microtiter plate format, although most of the techniques will apply to any format.

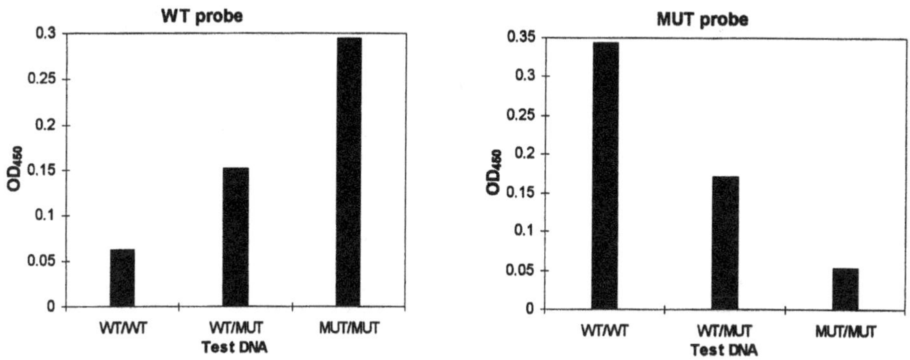

Fig. 1. IMBP Genotyping. Biotin-labeled probes were annealed to excess unlabeled PCR products amplified from genomic DNA of the following genotypes: (1) Homozygous wild-type (WT/WT); (2) Homozygous mutant (MUT/MUT) (Mutant sequence differs from wild-type by a single nucleotide substitution.); (3)Heterozygous wild-type/mutant (WT/MUT). Results are from IMBP 96-well plate assays.

The use of "probes" (labeled fragments amplified, in general, from cloned fragments and purified before use) *(8)* greatly enhances both the utility and the robustness of IMBP assays. With this method, biotin-labeled probes are generated from plasmid templates containing target wild-type (WT) or mutant (MUT) inserts by PCR amplification using one biotin-labeled and one unlabeled primer. The target region of test genomic DNA is amplified using unlabeled primers, portions of the resulting amplicons mixed separately with WT and MUT probes, and the samples hybridized by heat denaturation and reannealing. The hybridized fragments are added to IMBP plates. Test fragments that are identical to the probe will anneal to form perfectly matched duplexes (homoduplexes); these will not be retained by IMBP. Test fragments that are not identical to the probe sequence will anneal to form duplexes with mismatches (heteroduplexes) and will be retained by IMBP. For example, homoduplexes from WT/WT amplicons annealed to WT probe will not be bound by IMBP, whereas heteroduplexes from MUT/MUT amplicons annealed to WT probe will be bound. Annealing either probe with WT/MUT amplicons will produce an equal mixture of homoduplexes and heteroduplexes, which will give an intermediate signal in IMBP assays. Sample data are presented in **Fig. 1**. This chapter describes the probe method in detail. (For a discussion of the labeled primer method, *see* **Note 1**.) In addition to mutation/SNP detection, IMBP plates can be used for genotyping, PCR fidelity optimization, and rare sequence isolation.

2. Materials

2.1. IMBP Plates

IMBP plates and control oligonucleotides are obtained from Gene Check, Inc. (Fort Collins, CO).

2.2. Reagents

All reagents are from Sigma (St. Louis, MO) except where noted.

1. DNA amplification: AmpliTaq, Stoffel fragment (PE Applied Biosystems, Foster City, CA); dNTPs; amplification buffer (10 m*M* Tris-HCl, pH 8.3, 3.5 m*M* $MgCl_2$, 25 m*M* KCl, 10% glycerol); synthetic oligonucleotide primers (Macromolecular Resources, Fort Collins, CO). Amplification conditions should be optimized for each application or polymerase (*see* **Note 2**).
2. Reaction buffer: 20 m*M* Tris-HCl, pH 7.5, 5 m*M* $MgCl_2$, 0.1 m*M* dithiothreitol (DTT), 0.01 m*M* ethylenediaminetetraacetic acid (EDTA).
3. Hybridization buffer: 10 m*M* NaCl in Reaction Buffer.
4. Wash buffer: 0.05% Tween-20 in reaction buffer.
5. Streptavidin conjugated horseradish peroxidase (HRP; Pierce, Rockford, IL).
6. HRP colorimetric substrate: Tetramethylbenzine (TMB).
7. Stop acid: 0.5 *M* H_2SO_4.
8. Reagents for ligation of PCR products to plasmid vectors (e.g., TA cloning kit, InVitrogen, Carlsbad, CA).
9. PCR purification columns (e.g., QIAquick, Qiagen, Valencia, CA).

2.3. Apparatus

1. Thermocycler.
2. Plate reader with appropriate filters.

3. Methods

3.1. Probe Production

IMBP probe preparations are synthesized using one biotin-labeled and one unlabeled primer. Generally, a number of 100 µL PCR reactions are performed to generate large lots of probe. After amplification, the amplicons are pooled and a sample run in polyacrylamide gel electrophoresis to monitor quality and yield. The probe preparations are purified using glass matrix spin columns (QIAquick, Qiagen) to remove any unincorporated biotinylated primers, quantitated by OD_{260} and diluted to 1 ng/uL immediately prior to use.

3.1.1. Primer Orientation

When designing PCR primers for probe preparation, it is desirable to select the labeled primers to allow the best recognized mismatches (e.g., G:T) to be in labeled heteroduplexes when annealed with unlabeled test DNA. For example, for the sequences below, a biotin-labeled forward primer should be used to label the WT fragment (G-strand labeled) and a biotin-labeled reverse primer should be used to label the MUT fragment (T-strand labeled) (*see* **Note 3**).

WT

5'-------cctGca--------3'

3'-------ggaCgt---------5'

MUT

5'-------cctAca---------3'

3'------ggaTgt--------5'

Mixing WT probe with unlabeled mutant amplicon, denaturing and reannealing will produce the following mismatch containing heteroduplex:

bio*5'--------cctGca----------3' WT probe

3'--------ggaTgt----------5' MUT strand

Also present will be G:C homoduplexes, which will not bind to IMBP, and unlabeled C:A containing heteroduplexes that bind weakly to IMBP (compared to G:T containing heteroduplexes).

Similarly, mixing MUT probe with unlabeled WT amplicon, denaturing and reannealing will produce the following mismatch containing heteroduplex:

5'-------cctGca---------3' WT strand

3'-------ggaTgt---------5'*bio MUT probe

Also present will be A:T homoduplexes, which will not bind to IMBP, and unlabeled A:C containing heteroduplexes that bind weakly to IMBP (compared to G:T containing heteroduplexes).

3.1.2. Probe Template

Probe templates are plasmids containing WT or MUT inserts. These are constructed by ligating PCR products into plasmid vectors (e.g., TA cloning, InVitrogen). Sequencing is used to confirm the identity of the construct. Immediately prior to amplification, plasmid templates are diluted to 1 ng/µL and 1 µL is added to each 100-µL PCR reaction. We have successfully used genomic DNA as probe templates, although background signals are generally much higher. The advantage of cloned fragments for probe preparation is primarily the extremely high target to nontarget ratio for PCR (relative to genomic DNA template) and the consequent reduction in mispriming.

3.1.3. Probe Amplification Conditions

All amplification conditions, particularly annealing temperatures, should be individually determined for each primer set. Annealing temperatures for probe amplification are generally as high as possible, i.e., at or near the theoretical G:C + A:T melting temperature.

3.1.4. Probe Concentration

Prior to being used in an IMBP assay, the biotin-labeled probes are diluted to 1 ng/μL in hybridization buffer. 10 μL (10 ng) of diluted probe is mixed with test DNA. This quantity and concentration of probe works well for 100–400 bp fragments.

3.2. Test DNA Amplification

All amplification conditions, particularly annealing temperatures, should be individually determined for each primer set. After the PCR conditions (buffers, annealing temperature, fidelity, and so on) for a particular target are optimized, amplified test samples are not generally quantitated or checked by gel electrophoresis.

3.3. Probe/Test DNA Hybridization

1. Dilute the WT and MUT probes to 1 ng/μL with hybridization buffer, add 10 μL of the probe dilution to each tube or well.
2. Add 5–15 μL of test amplicon to tubes or wells containing the WT and MUT probes. [In general, the ratio of probe to test should be on the order of 1:10 (0.1 picomole probe : 1 picomole test DNA amplicon). This usually achieved by using 5–15 μL of test DNA amplicon to mix with 10 ng of probe.]
3. Denature and reanneal the samples by heating to 94°C for 3 min; annealing at 75°C for 30 min; and quickly cooling to 4°C for at least 15 min. [This temperature profile has worked well for most fragments; however, each application may need to be optimized.]
4. Add 10 μL of wash buffer. This step simply increases the volume to 30–35 μL, which facilitates transferring samples to the IMBP plate.

3.4. IMBP Assay (See Note 4)

1. Remove storage buffer from the wells by rinsing with 150 μL wash buffer and shaking off the liquid. Rinse four more times. After the last rinse, rap the inverted plate on a pile of laboratory tissues to remove any remaining liquid. Examine each well to ensure that no wash buffer remains. To avoid background signals, do not allow wells to dry once they have been hydrated and avoid the formation of bubbles in the wells. All incubations are at room temperature.
2. Add DNA samples (20–40 μL per well). Cover the plate and incubate for 15–30 min. Whereas DNA is binding, make up Streptavidin-HRP solution by adding 5 μL conjugate to 10 mL wash buffer (sufficient for one plate).
3. Add 150 μL wash buffer to DNA containing wells and shake out liquid. Rinse four additional times. After the last rinse, rap the plate on tissues and check for any residual liquid.
4. Add 100 μL diluted Streptavidin-HRP solution to each well. Cover the plate and incubate for 15–20 min.

5. Shake off Steptavidin-HRP and rinse the wells five times with wash buffer (150 µL). Again, be certain that all wash is removed from the wells using the rapping technique.
6. Add 80 µL TMB to each well. Allow 10–30 min for blue color development. Avoid allowing color to develop for so long that precipitate develops. The blue color can be read at 370 or 655 nm. The sensitivity of the assay is enhanced by adding 80 µL of stop acid (0.5 *M* H_2SO_4) to each well and reading the resulting yellow color at 450 nm. Wipe the bottom of the plate before reading.

4. Notes

1. Probes vs labeled primers. The IMBP method detailed in this paper utilizes probes for mutation/SNP detection. It is possible to use biotin-labeled primers and directly amplify genomic test DNA to detect heterozygous mutations/SNPs *(7)*. The major advantage of the labeled primer method is that it allows detection of heterozygous sequences without cloning or probe preparation. In addition, it allows the use of only a single biotin-labeled primer, because both mismatches of a WT/MUT pairing will be labeled from a genomic DNA template. Therefore, the labeled primer method is well suited to rapid scanning applications in which mutations and SNPs occur predominantly in heterozygous form and where many different DNA fragments or regions need to be examined in a relatively small number of samples. The disadvantages of the labeled primer method are as follows.
   a. Background signals will be higher. Amplifying genomic DNA with labeled primers allows for the production of labeled mispriming products. Labeled single-stranded DNA fragments may adopt secondary structures containing substrates for IMBP binding (mispaired or unpaired bases). Further, it may be necessary to reduce stringency of PCR amplification conditions (annealing of primer) when using biotin-labeled primers, which can increase the extent of mispriming.
   b. Quantitation of PCR products is necessary for each sample to obtain comparable signals. This quantitation generally requires gel electrophoresis.
   c. There is no internal control for PCR failure. With the labeled primer method, PCR failure will be scored as homozygous. Thus, gel electrophoresis is generally required to confirm successful amplification. With the probe method, failed PCR will score as homozygous wild type with the wild-type probe and homozygous mutant with the mutant probe.
   d. There is no way to distinguish WT and MUT homozygotes—both will score as negatives.
2. PCR conditions. The real key to successful IMBP assays is in providing the assay with good DNA. Perhaps the major source of background-generating errors is the amplification step. PCR errors are of two kinds: (1) mispriming, which is primarily a problem in labeled primer formats (*see* above); and (2) misincorporation. PCR fragments containing misincorporation errors will form mismatch containing heteroduplexes when they are paired with probes and these heteroduplexes will generate background signals. Control of misincorporation depends on several factors:

a. Fragment length. The longer a PCR fragment, the more likely it is to contain a misincorporation error. Fragment length of less than 400 bp is best suited to IMBP assays, although longer fragments have been used successfully.
b. Cycle number. The more cycles a fragment is amplified, the higher the background. Misincorporation errors are created at a roughly constant rate per nucleotide duplicated. However, once created, an error is duplicated in each succeeding round of amplification, such that errors accumulate during PCR.
c. Polymerase. High-fidelity synthesis is essential to low-background IMBP assays. For single-enzyme amplifications, the highest fidelity is obtained with proofreading polymerases such as Pfu (Stratagene, LaJolla, CA) *(8)*. Unfortunately, proofreading polymerases give relatively poor yields and are frequently difficult to use because they digest primers and products, as well as polymerase errors. The IMBP assay protocol presented in this paper uses the Stoffel fragment of *Taq*. The fidelity of this polymerase and others like it (e.g., Klen*Taq*, Ab Peptides, St. Louis, MO) comes from the fact that they are extremely nonprocessive and most likely do not extend mispaired termini. Perhaps the best solution is a mixture of a Stoffel-like polymerase with a proofreading polymerase. Such mixtures are available commercially, work well in IMBP assays, but are currently expensive.
d. Template. Use the lowest amount of template that still gives reasonable yield. Too much template can reduce signal (*see* **Note 3**).
e. Cycling temperatures. Cycling conditions should include annealing temperatures as high as possible to minimize mispriming. Hot start may also be beneficial, particularly when using labeled primers.
f. Buffer conditions. Buffer conditions can have an enormous impact on PCR fidelity (**Fig. 2**) *(9)*. It is important to remember that, prior to IMBP assays, polymerase fidelity was extremely time consuming to measure. Therefore, and because in many applications yield is more important than fidelity, many buffers sold with enzymes are optimized for yield rather than for fidelity. Taking the time to optimize PCR conditions before IMBP assays can greatly increase the likelihood of successful assays.
g. Optimizing PCR. Optimizing PCR buffer conditions is simple with IMBP assays and using labeled primers is the simplest way to use the IMBP assays for optimization *(9)*. Use a single biotinylated primer to amplify the fragment of interest from a known homozygous sample and compare the signal obtained from different buffers in IMBP assays. Run gels to confirm the presence of amplicon. The lowest signal is the highest fidelity. It may be necessary to increase cycle number to accentuate the fidelity differences between various buffers. Stratagene's Optiprime Kit provides a useful range of buffers and additives to use in an optimization protocol.

3. Competing DNA. Unlabeled, mismatch-containing DNA can compete with labeled DNA for binding to IMBP plates and give lowered signals. Therefore, as dis-

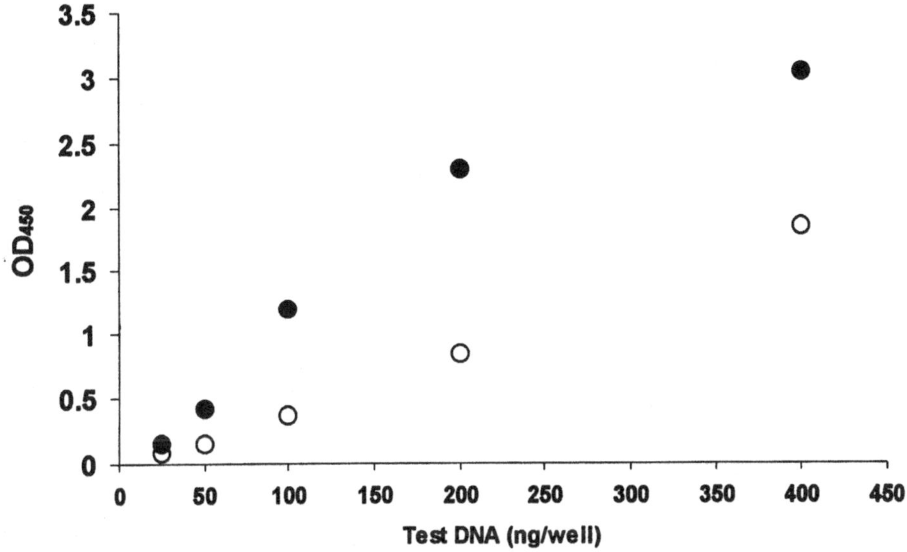

Fig. 2. Effects of buffer on PCR fidelity. Two buffers (10 m*M* Tris (pH 8.3) 3.5 m*M* $MgCl_2$, 25 m*M* KCl (l) or 10 m*M* Tris (pH 8.8) 1.5 m*M* $MgCl_2$, 25 m*M* KCl (*m*) were tested using Pfu DNA polymerase (Stratagene) and biotin-labeled primers to amplify a 172-bp region of the human *p53* gene. The resulting amplicons were denatured and reannealed and various concentrations added to wells of an IMBP plate. Following streptavidin-HRP binding and colormetric detection, the comparative difference between the buffers were measured. With IMBP assays, the lower the rate of misincorporation, the lower the IMBP signal; therefore, the 1.5 m*M* $MgCl_2$ (pH 8.8) buffer gave higher PCR fidelity.

cussed above, it is important to select probes to label the heteroduplex with the most readily bound mismatch. In addition, it is important to minimize the amount of genomic DNA template going into PCR. Denatured genomic DNA is an excellent competitor in IMBP assays.

4. Enrichment. IMBP plates can be used to isolate rare sequences from DNA preparations. DNA can released from the IMBP with 0.5 *M* NaCl. Released DNA can be cloned (clones will be "mixed," i.e., will contain both MUT and WT sequences, because all fragments that bind to IMBP must be heteroduplexes) or amplified and sequenced. A rare sequence will be revealed as a heterozygote at the site of the sequence difference. The success of enrichment depends on minimizing PCR errors (see above) because heteroduplexes formed by error containing strands will also be bound by IMBP and will be cloned or amplified in competition with the rare sequence strands of interest.

References

1. Modrich, P. and Lahue, R. (1996) Mismatch repair in replication fidelity, genetic recombination, and cancer biology. *Annu. Rev. Biochem.* **65,** 101–133.
2. Dohet, C., Wagner, R., and Radman, M. (1985) Repair of defined single-base-pair mismatches in *Escherichia coli. Proc. Natl. Acad. Sci. USA* **82,** 503–505.
3. Su, S.-S., Lahue, R. S., Au, K. G., and Modrich, P. (1988) Mispair specificity of methyl-directed DNA mismatch correction in vitro. *J. Biol. Chem.* **263,** 6829–6835.
4. Fazakerley, G. V., Quignard, E., Woisard, A., et al. (1986) Structures of mismatched basepairs in DNA and their recognition by the *E. coli* mismatch repair system. *EMBO J.* **5,** 3697–3703.
5. Jones, M., Wagner, R., and Radman, M. (1987) Repair of a mismatch is influenced by the base composition of the surrounding nucleotide sequence. *Genetics* **115,** 605–610.
6. Parker, B. O. and Marinus, M. G. (1992) Repair of DNA heteroduplexes containing small heterologous sequences in *Escherichia coli. Proc. Natl. Acad. Sci. USA* **89,** 1730–1734.
7. Wagner, R., Debbie, P., and Radman, M. (1995) Mutation detection using immobilized mismatch binding protein (MutS). *Nucleic Acids Res.* **23,** 3944–3948.
8. Debbie, P., Young, K., Pooler, L., et al. (1997) Allele identification using immobilized mismatch binding protein: detection and identification of antibiotic resistant bacteria and determination of sheep susceptibility to scrapie. *Nucleic Acids Res.* **25,** 4825–4829.
9. Wagner, R. and Dean, A. (1999) The use of immobilized mismatch binding protein for the optimization of PCR fidelity, in *PCR Applications* (Innis, M., Gelfand, M., and Sninsky, J., eds.), Academic, New York, pp. 95–104.

14

Use of Uracil DNA Glycosylase in the Detection of Known DNA Mutations and Polymorphisms

Glycosylase-Mediated Polymorphism Detection (GMPD-Check)

Pat Vaughan

1. Introduction

Detection of known mutations and polymorphisms at specific locations in DNA is now the basis for genetic diagnosis, molecular genotyping, molecular epidemiology, and rapid identification of various traits in eukaryotic and prokaryotic organisms. An optimum system for sequence variation detection needs to offer specificity, sensitivity, flexibility, as well as ease of use and allow rapid and high throughput of samples.

The method described here, and represented diagrammatically in **Fig. 1.**, is based on the fact that the normal four nucleotides that constitute DNA can be specifically and individually replaced completely by related modified nucleotides. These modified bases are then susceptible to the action of specific DNA glycosylases that excise the modified base from the DNA and the resultant abasic site can be subsequently cleaved either enzymatically or chemically *(1,2)*. These modified bases can be introduced into DNA by chemical treatment of the DNA, or more commonly by the incorporation of the corresponding modified nucleotides into DNA during polymerase chain reaction (PCR) amplification.

The DNA glycosylase described here is uracil-DNA-glycosylase (UDG) which is a highly specific DNA repair enzyme that cleaves the N-glycosidic bond between the base uracil and the sugar deoxyribose in a DNA molecule and generates an apyrimidinic (AP) site *(3,4)*. The corresponding modified nucleotide which is incorporated into the DNA, is dUMP. This is usually achieved by supplying dUTP to the PCR amplification reaction.

From: *Methods in Molecular Biology, vol. 152: DNA Repair Protocols: Prokaryotic Systems*
Edited by: P. Vaughan © Humana Press Inc., Totowa, NJ

The sample DNA is amplified using three of the four normal precursor nucleotides (dGTP, dATP, and dCTP) and one modified precursor nucleotide, dUTP, which completely replaces dTTP (*see* **Notes 1** and **2**). One primer used in the amplification is normally labeled and is referred to as the diagnostic primer (*see* **Fig. 1** and **Notes 3** and **4**). The modified base, in this case uracil, is then excised with Uracil DNA glycosylase, which generates an AP site. The phosphate linkages at the AP sites are then cleaved with 5' or 3' specific agents (*see* **Note 5**). Cleavage of the amplified DNA will yield a DNA fragment of specific length, which is a measure of the distance between the 5' end of the diagnostic primer and the first site of incorporation of dUMP upon extension of the 3' end of that primer. The presence of a mutation results in the appearance of a cleavage fragment following denaturing polyacrylamide gel electrophoresis (PAGE) that may be larger or smaller than the wild-type fragment, depending on whether the mutation results in the loss or gain of a dUMP incorporation site, respectively. A heterozygote sample will result in the observation of the two fragments.

The glycosylase-mediated polymorphism detection (GMPD-check) method has many attractive features that offer significant advantages over several current technologies utilized for detection of known mutations. In particular, GMPD offers the specificity of direct DNA sequencing, but is a much simpler process with almost all mutations detected using a single enzyme (*see* **Notes 1** and **2**). The method does not rely on heteroduplex formation, but works equally well on homozygous and heterozygous samples. The annealing of primers is not dependent on stringent or differential hybridization conditions. Overall, only sequence information of 20–25 nucleotides on either side of the mutation site is required in order to design appropriate primers for the amplification process. The process can be performed directly on any DNA sample and the process is robust and requires relatively few manipulations. Diagnostic and nondiagnostic primers can be designed to allow multiplexing of polymorphism detection (*see* **Notes 6–8**). Solid-phase GMPD using one of the many applicable detection systems avoids the use of time-consuming gel electrophoresis and allows rapid and high throughput sample analysis (*see* **Note 9**).

2. Materials

2.1. 5' End Labeling of Oligonucleotides with ^{32}P

1. T4 polynucleotide kinase (NEB Biolabs) 1 U/μL, supplied with 10X T4 PNK reaction buffer: 70 m*M* Tris-HCl, pH 7.6, 10 m*M* $MgCl_2$, 5 m*M* dithiothreitol (DTT).
2. $\gamma\,^{32}P$ ATP, specific activity 3000 Ci/mmol.
3. 4 *M* ammonium acetate.
4. Absolute ethanol.
5. CentriSpin-10 column (Princeton Separations, supplied by Genosys, UK).

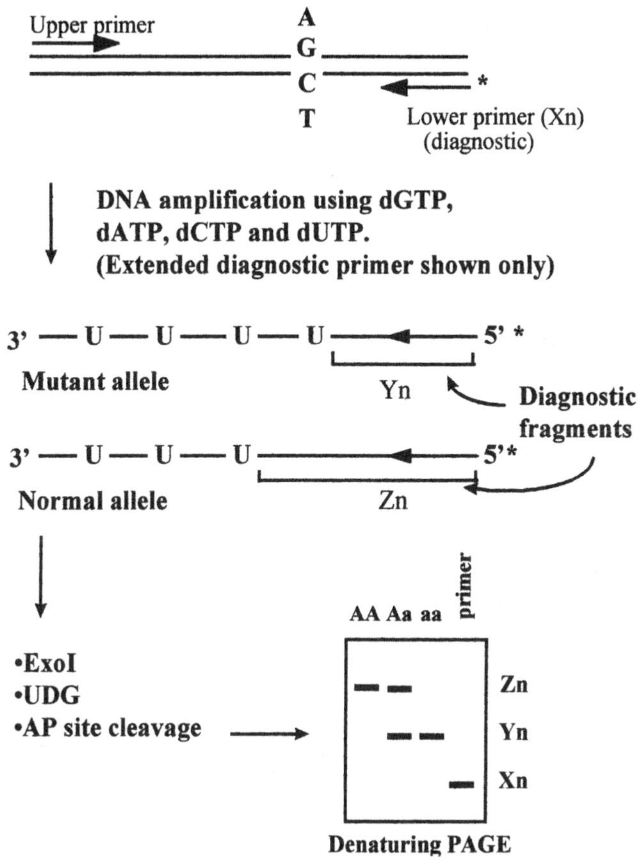

Fig. 1. Schematic diagram of GMPD-check. For visual simplicity, only extension of the diagnostic primer following PCR amplification is shown and the extension of the nondiagnostic primer is not shown in the diagram. The DNA fragment shown here is an arbitrary piece of DNA. The mutation site is depicted by a G to A mutation and C to T mutation on the upper and lower strands, respectively. "Normal" thymine positions in the arbitrary DNA fragment are depicted by "U" in the lower strand following extension of the diagnostic primer during PCR amplification in the presence of dUTP, which completely replaced dTTP.

2.2. Amplification of Sample DNA

1. dNTPs, including dUTP.
2. PCR reagents: commercially available from many sources and used as per supplier's instructions.
3. α^{32}P dCTP (3000 Ci/mmol).
4. Wild-type and mutant template DNA and oligonucleotide primers.

2.3. Glycosylase-Mediated Cleavage of Amplified DNA

1. Exonuclease I (diluted to 0.5 U/μL) (Amersham Life Sciences, UK).
2. Uracil DNA glycosylase (UDG) (diluted to 0.5 U/μL) (NEB Biolabs).
3. Dilution buffer: 70 m*M* HEPES-KOH, pH 8.0, 1 m*M* DTT, 1 m*M* ethylenediaminetetraacetic acid (EDTA), 50% glycerol.
4. NaOH, 5 *M* stock stored in plastic (not glass) container. Make up fresh every 5–6 wk.
5. Tris base (Sigma).
6. Formamide loading dye: 95% v/v formamide, 0.5% w/v bromophenol blue, 0.5% xylene cyanol.
7. Formamide loading dye for Alf Express: 95% v/v formamide, 0.8% w/v dextran blue (*see* **Note 10**).

2.4. Polyacrylamide Gel Electrophoresis

1. Urea.
2. 40% w/v acrylamide:*bis*-acrylamide (29:1).
3. TBE: 90 m*M* Tris base, 90 m*M* boric acid, 2 m*M* EDTA.
4. N,N,N',N'-tetramethylethylenediamine (TEMED).
5. 10% w/v ammonium persulphate. Make up fresh each time, or store in small single-use aliquots at –20°C and thaw immediately before use.

3. Methods

3.1. 5' End Labeling of Oligonucleotide Primers with ^{32}P

1. Place 50–60 pmol (approx 400 ng) of the diagnostic primer in 1.5-mL Eppendorf tube.
2. Add 5 μL γ ^{32}P ATP (3000 Ci/mmol), 2.5 μL 10X T4 PNK reaction buffer and bring volume to 24 μL with sterile H_2O (MilliQ grade).
3. Add 1 μL T4 polynucleotide kinase (1 U/μL) and incubate reaction for 30 min at 37°C.
4. To precipitate the labeled oligonucleotide and remove unused γ ^{32}P ATP, add 25 μL 4 *M* ammonium acetate and 100 μL ethanol and leave at –20°C for at least 2 h Alternatively, the oligonucleotide can be purified by using centrifugation through a gel-filtration column (*see* **Note 11**).

3.2. Amplification of Sample DNA

When using labeled oligonucloetides (*see* **Note 12**), 6 pmoles of upper and lower primer are used to amplify the DNA fragment of interest in a PCR in the presence of 0.2 m*M* dATP, dCTP, and dGTP and 0.2 m*M* dUTP, in a total volume of 20 μL. Normally, 100–200 ng genomic DNA sample serves as template for this reaction. It is recommended that a PCR of a normal or wild-type sample (and cleavage of same) is always included in each analysis for comparison purposes.

When using unlabeled oligonucleotides in the presence of labeled dNTP, 3 pmoles of upper and lower primer are used to amplify the DNA fragment of interest in a PCR in the presence of 0.2 m*M* dATP, dGTP, and dUTP, 0.02 m*M* dCTP, and 0.2 µL α^{32}PdCTP in a total volume of 20 µL (*see* **Note 13**).

3.3. Glycosylase-Mediated Cleavage of Amplified DNA

1. 5 µL of PCR reaction is placed in 0.5 µL Eppendorf tube.
2. Add 1 µL exonuclease I (0.5 U/µL) and incubate at 37°C for 30 min. Enzyme is diluted using dilution buffer.
3. The exonuclease I is inactivated by incubating the reaction at 80°C for 15 min.
4. Following heat inactivation, quick centrifuge tube to return all solution to end of tube and put tube on ice.
5. Add 1 µL uracil DNA-glycosylase (0.5 U/µL) and incubate at 37°C for 30 min.
6. Add 1 µL 0.4 *M* NaOH (final concentration of 0.05 *M* NaOH) and heat the reaction for 15 min at 95°C.
7. Add 1 µL 0.27 *M* Tris base (final concentration of 0.03 *M* Tris base) to neutralize the reaction.
8. Add 5 µL formamide loading dye to the sample.
9. Store at –20°C until ready to carry out gel analysis.
10. Heat to 90°C for 5 min prior to loading on polyacrylamide gel.
11. For samples containing fluorescent-labeled primers, store at –20°C without adding loading dye. Add dye immediately prior to heating and loading on gel. Prolonged storage of the digest in loading buffer causes degradation of fragments and/or label.

3.4. PAGE

3.4.1. Manual PAGE and Autoradiography

1. Set up 20% denaturing polyacrylamide gel, containing 7 *M* urea and 1X TBE.
2. Wash out wells well with running buffer (1X TBE).
3. Load 5 µL of each sample (*see* **Note 14**).
4. Load labeled primer and uncleaved PCR product for use as size markers during fragment analysis.
5. Electrophorese at 500 V for 3–4 h or until bromophenol blue dye has reached the end of the gel.
6. Expose the gel to X-ray film for several hours (*see* **Note 15**).

3.4.2. PAGE Using Alf Express

Because the fragments produced by GMPD-check are relatively small, a high percentage acrylamide gel using the short Alf Express (Pharmacia) plates will suffice to separate the cleaved fragments. This also reduces the time required to ensure efficient fragment separation.

1. Make up a 15% denaturing PAG solution consisting of 21 g urea (7 *M*), 1X TBE (*see* **Note 16**) and 18.75 mL of 40% acrylamide stock. Bring volume to 50 mL with MilliQ grade H_2O. Solution may be heated to help dissolve urea.

2. Filter solution through at least a 0.45-μm filter.
3. Wash and set up gel plates according to manufacturer's instructions.
4. Allow gel solution to return to room temperature and add 35 μL TEMED and 350 μL 10% APS to the gel solution and gently mix.
5. Pour gel according to manufacturer's instructions, taking care to avoid the incorporation of air bubbles. Allow gel to polymerize fully for at least 2 h.
6. Set up gel in automated sequencer, wash out wells well with running buffer (1X TBE) and load 4 μL of sample per well.
7. Load labeled primer and uncleaved PCR product for use as size markers during fragment analysis.
8. Run gel at 55°C and at 500 V (constant) for 1.5–2 h (*see* **Note 17**).

3.6. Fragment Analysis

1. Determine the position of the labeled primer on the gel.
2. All detectable cleaved fragments will be similar or greater in size than this band.
3. Determine the position of the cleavage fragment for the normal or wild-type sample.
4. If the base change at the mutation site involves a loss of a T residue, then a band larger in size than the normal band will be detected, corresponding in size to the position of the next T residue 3' of the mutation site.
5. If the base change at the mutation site involves a gain of a T residue, then a band smaller in size than the normal band will be detected, corresponding in size to the position of the mutant T residue 5' of the first normal T position.
6. If a suitable size marker (labeled) is used during PAGE, then band size can be estimated by comparison to this marker.

4. Notes

1. GMPD-check using UDG and dUTP incorporation detects all deletion and insertion mutations. As 10 out of the 12 base substitution mutations involve the gain or loss of A/T or T/A base pairs (i.e., G>A, G>T, A>G, A>T, A>C, T>G, T>A, T>C, C>A, and C>T), 83% of mutations can be detected with UDG and dUTP. In reality this figure is closer to 95% or greater of all point mutations because most disease-causing mutations are C to T (or G to A) transitions.
2. The two base changes not detected with UDG and dUTP incorporation, i.e. G to C and C to G occur at a low frequency relative to other base substitution mutations. These changes can be detected by using a modified precursor nucleotide that substitutes for dGTP or dCTP and a corresponding DNA glycosylase.
3. Diagnostic primers are designed so that during extension, the position of the first uracil incorporated into the extended primers differs depending on whether a mutation is present or absent. Importantly, there must be no uracil incorporation site between the 3' end of the diagnostic primer and the mutation site.
4. The oligonucleotide primers contain dG, dA, dT, and dC and consequently are resistant to cleavage by UDG.

5. AP sites can be cleaved at the phosphate linkage on the 5' and / or 3' side of the ribose ring. Heating of the DNA at 95°C in the presence of alkali causes chemical cleavage at the 5' side of the AP site and leaves a phosphate group at the 3' terminus of the diagnostic fragment. AP endonuclease IV enzymatically cleaves at the 5' side of the AP site and leaves a 3' OH terminus on the diagnostic fragment *(5)*. However, cleavage of AP sites by heating at 95°C at neutral pH results in chemical cleavage at the 3' side of the AP site, which leaves a deoxyribose phosphate moiety at the 3' terminus of the diagnostic fragment and a low level of cleavage at the 5' side leaving a 3' phosphate group on the diagnostic fragment. This results in the diagnostic fragment migrating as a doublet following denaturing PAGE. The presence of this 3' deoxyribose phosphate group distinguishes the diagnostic fragment from the unused primer where the diagnostic primer is designed such that its 3' terminus binds at the nucleotide directly 5' adjacent to the mutation site *(6)*. An AP lyase enzyme can also be used in place of the heat treatment because lyases also cleave at the 3' side of the AP site and result in a 3' deoxyribose phosphate terminus on the diagnostic fragment *(7)*.
6. Use of differentially labeled diagnostic primers allows one to analyse multiple different mutations in each gel lane, allowing high sample throughput.
7. Careful design of diagnostic primers of differing lengths can also allow multiplexing of samples/mutations per gel lane, allowing for higher throughput fragment separation.
8. Use mismatches and inosine residues in the nondiagnostic primer in order to change the size of the normal allele will facilitate multiplex detection of mutations *(8)*. Inosines in the template strand usually direct the incorporation of cytosine residues into the newly synthesised strand *(9,10)*. Some or all of the adenine residues within the primer are replaced with inosine residues or with a mismatched base. Replacement of the penultimate 3' adenine base in the nondiagnostic primer (upper primer) with guanine or inosine increases the length of the normal allele following glycosylase mediated cleavage of the amplified product. Replacement of all the adenines in the nondiagnostic primer with inosines yields a labeled normal allele that is devoid of uracil bases and resistant to cleavage mediated by UDG.
9. Replacement of adenines with inosines in a primer exclude incorporation of uracil residues in the newly synthesized DNA complementary to the primer strand. This facilitates solid phase GMPD because UDG resistant or sensitive strands of DNA can be amplified dependent on presence or absence of the incorporation of a uracil residue at the mutation site *(8)*.
10. When using the ALF Express for fragment separation and analysis (and/or use of Cy5 label on any fluorescence detection equipment), one must use a modified formamide loading dye containing dextran blue in place of bromophenol blue and xylene cyanol, because bromophenol blue and xylene cyanol fluoresce with 635 nm excitation and, hence, interfere with the detection of the Cy5 fluorescent label.

11. Following 5' end labeling of the oligonucleotide, unincorporated γ^{32}PATP can be removed by ethanol precipitation or by using homemade gel filtration columns or any of the commercially available oligonucleotide purification kits (e.g., CentriSpin-10 columns, Princeton Separations, supplied by Genosys, UK).
12. The diagnostic oligonucleotide can be labeled with a variety of different labels. Radioactive label, such as ^{32}P, is detected using autoradiography and/or phosphor-imaging technology. The use of fluorescently labeled oligonucleotides is also possible with many different fluorescent labels currently available, e.g., the label referred to in this chapter is Cy5, which is used in conjunction with the Alf Express automated sequencer. Biotin labeling of oligonucleotides and detection via labeled avidin is also an option.
13. DNA can be labeled for GMPD analysis by incorporation of a labeled dNTP during PCR amplification in place of using a 5'end-labeled oligonucleotide. In the more common scenario of using UDG and dUTP incorporation, α^{32}PdGTP, α^{32}PdATP, or α^{32}PdCTP can be used as the labeled nucleotide and α^{32}PdTTP may not be used.
14. Normally, one may load 5 µL of sample per well. However, this may be altered, depending on how well the oligonucleotide primer was labeled with ^{32}P and how long one is willing to leave the radioactive gel exposed to the X-ray film.
15. A good starting point for X-ray film exposure is overnight exposure at –20°C.
16. Do not autoclave TBE for use with the Alf Express.
17. The actual running time of the gel is determined by the size of the expected cleavage fragments. The progress of the fragments can be estimated by assuming bromophenol blue runs approx as an 8 mer and xylene cyanol as a 28 mer in a 20% denaturing PAG.

References

1. Lindahl, T. and Andersson, A. (1972) Rate of chain breakage at apurunic sites in double stranded deoxyribonucleic acid. *Biochemistry* B, 3618–3623.
2. Price, A. and Lindahl, T. (1991) Enzymatic release of 5'-terminal deoxyribose phosphate residues from damaged DNA in human cells. *Biochemistry* **30,** 8631–8637.
3. Lindahl, T., Ljungquist, S., Siegert, W., et al. (1977) DNA N-Glycosidases, properties of uracil-DNA glycosylase from *Escherichia coli*. *J. Biol. Chem.* **252,** 3286–3294.
4. Duncan, B. (1981) *The Enzymes*, vol. 14 (Boyer, P., ed.), Academic, Orlando, FL, pp. 565–586.
5. Doetsch, P. W. and Cunningham, R. P. (1990) The enzymology of apurinic/apyrimidinic endonucleases. *Mutat. Res.* **236,** 173–201.
6. Vaughan, P. and McCarthy, T. V.,(1999) Glycosylase mediated polymorphism detection (GMPD): a novel process for genetic analysis. *Genet. Anal. Biomol. Eng.* **14,** 169–175.
7. Levin, J. D. and Demple, B. (1990) Analysis of class II (hydrolytic) and class I (B-lyase) apurinic/apyrimidinic endonucleases with a synthetic DNA substrate. *Nucleic Acids Res.* **18,** 5069–5075.

8. Vaughan, P. and McCarthy, T. V. (1998) A novel process for mutation detection using uracil DNA-glycosylase. *Nucleic Acids Res.* **26,** 810–815.
9. Martin, F. H. and Castro, M. M. (1985) Base pairing involving deoxyinosine: implications for probe design. *Nucleic Acids Res.* **13,** 8927–8938.
10. Case-Green, S. C. and Southern, E. M. (1993) Studies on the base pairing properties of deoxyinosine by solid phase hybridisation to oligonucleotides. *Nucleic Acid Res.* **22,** 131–136.

15

Use of Uracil DNA Glycosylase in Scanning for Unknown DNA Mutations and Polymorphisms

Glycosylase-Mediated Polymorphism Detection (GMPD-Scan)

Pat Vaughan

1. Introduction

Detection of unknown mutations and polymorphisms is a central ingredient in disease gene discovery and genetic diagnosis in addition to polymorphism discovery and genotyping programs. An optimum system for detection of sequence variation needs to offer specificity, sensitivity, flexibility, as well as ease of use.

The GMPD scan described here and represented diagrammatically in **Fig. 1**, is based on the fact that the normal four nucleotides that constitute DNA can be partially replaced by specific and related modified nucleotides *(1)*. These modified bases are then susceptible to the action of specific DNA glycosylases that excise the modified base from the DNA and the resultant abasic site can be enzymatically or chemically cleaved *(2,3)*. These modified bases can be introduced into DNA by chemical treatment of the DNA, or more commonly by the incorporation of the corresponding modified nucleotides into DNA during polymerase chain reaction (PCR) amplification.

Briefly, the sample DNA is amplified using the four normal precursor nucleotides (dGTP, dATP, dTTP, and dCTP) and one modified precursor nucleotide. The essence of this technique, as mentioned above, is that the modified precursor nucleotide partially replaces one of the normal dNTPs. The modified precursor nucleotide described here is dUTP, which can substitute for dTTP in a DNA polymerization reaction. A low level of dUMP is introduced randomly into the amplified target gene sequence during DNA amplification, carried out

From: *Methods in Molecular Biology, vol. 152: DNA Repair Protocols: Prokaryotic Systems*
Edited by: P. Vaughan © Humana Press Inc., Totowa, NJ

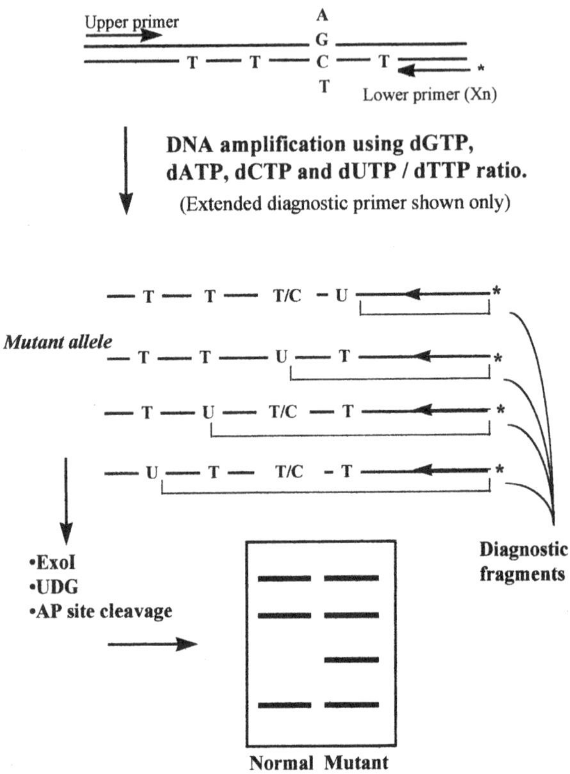

Fig. 1. Schematic diagram of GMPD-scan. For visual simplicity, only extension of the diagnostic primer following PCR amplification is shown and the extension of the nondiagnostic primer is not shown in the diagram. The DNA fragment shown here is an arbitrary piece of DNA. The mutation site is depicted by a G to A mutation and C to T mutation on the upper and lower strands, respectively. "Normal" thymine positions in the arbitrary DNA fragment in the lower strand are depicted by "T" and by "T" or "U" following extension of the diagnostic primer during PCR amplification in the presence of dUTP and dTTP.

in the presence of the four normal precursor nucleotides and dUTP. Therefore, the reaction contains both dTTP and dUTP. The ratio of the dUTP to dTTP is such that one U is incorporated per strand of amplified DNA (*see* **Note 1**). One primer is labeled prior to amplification so as to analyze one strand of the target sequence. Both strands must be analyzed to detect all possible mutations within that fragment. Both strands may be analyzed simultaneously by using differentially labeled primers (*see* **Note 2**). See schematic diagram in **Fig. 1**. The modified base, in this case uracil, is then excised with Uracil DNA glycosylase. Uracil-DNA-glycosylase (UDG) is a highly specific DNA repair enzyme that cleaves the N-glycosidic bond between the base uracil and the sugar deoxyribose in a DNA molecule and generates an apyrimidinic (AP) site *(4,5)*. The phosphate linkages at the AP sites are then cleaved with 5' or 3' specific agents (*see* **Note 3**). As the uracil is incorporated randomly, glycosylase-mediated cleavage of the amplified DNA will yield a population of DNA fragments of various lengths. The individual cleavage products are separated by gel electrophoresis and this produces a ladder of bands somewhat identical to the T lane of a sequencing gel. The number of bands indicates the number of Ts in the original target sequence. The appearance or disappearance of a cleavage fragment when compared to the known DNA sequence indicates the presence of a mutation. The location and sequence of the mutation can then be determined by size analysis of the normal and mutant fragments.

2. Materials

2.1. 5' End Labeling of Oligonucleotides with ^{32}P

1. T4 polynucleotide kinase (New England Biolabs) 1 U/µL, supplied with 10X T4 PNK reaction buffer: 70 m*M* Tris-HCl, pH 7.6, 10 m*M* $MgCl_2$, 5 m*M* dithiothreitol (DTT).
2. $\gamma\ ^{32}P$ ATP, specific activity 3000 Ci/mmol.
3. 4 *M* ammonium acetate.
4. Absolute ethanol.
5. CentriSpin-10 column (Princeton Separations, supplied by Genosys, UK).

2.2. Amplification of Sample DNA

1. Labeled and unlabeled oligonucleotides (*see* **Notes 2** and **4**).
2. dNTPs, including dUTP.
3. PCR reagents: commercially available from many sources and used as per supplier's instructions.
4. $\alpha^{32}P$ dCTP (3000 Ci/mmol)
5. Wild-type and mutant template DNA and oligonucleotide primers.

2.3. Glycosylase-Mediated Cleavage of Amplified DNA

1. Exonuclease I (diluted to 0.5 U/µL) (Amersham Pharmacia Biotech).
2. CentriSpin-10 column (Princeton Separations, supplied by Sigma Genosys) or S300HR spin columns (Amersham Pharmacia Biotech).
3. Uracil DNA glycosylase (UDG) (diluted to 0.5 U/µL) (New England Biolabs).
4. Dilution buffer: 70 m*M* HEPES-KOH, pH 8.0, 1 m*M* DTT, 1m*M* ethylenediaminetetraacetic acid (EDTA), 50% glycerol.
5. NaOH, 5 *M* Stock stored in plastic (not glass) container. Make up fresh every 5–6 wk.
6. Tris base (Sigma).
7. Formamide loading dye: 95% v/v formamide, 0.5% w/v bromophenol blue, 0.5% xylene cyanol.

8. Formamide loading dye for Alf Express: 95% v/v formamide, 0.8% w/v dextran blue (*see* **Note 5**).

2.4. Polyacrylamide Gel Electrophoresis (PAGE)

1. Urea
2. 40% w/v acrylamide:*bis*-acrylamide (29:1).
3. 40% w/v acrylamide:*bis*-acrylamide (99:1).
4. TBE: 90 m*M* Tris base, 90 m*M* boric acid, 2 m*M* EDTA (*see* **Note 6**).
5. N,N,N',N'-Tetramethylethylenediamine (TEMED).
6. 10% w/v ammonium persulfate.
7. Glycerol.

3. Methods

3.1. 5' End Labeling of Oligonucleotide Primers with ^{32}P

1. Place 50–60 pmoles (approx 400 ng) of the diagnostic primer in 1.5-mL Eppendorf tube.
2. Add 5 µL $\gamma\,^{32}P$ ATP (3000 Ci/mmol), 2.5 µL 10X T4 PNK reaction buffer and bring volume to 24 µL with sterile H_2O (MilliQ grade).
3. Add 1 µL T4 polynucleotide kinase (1 U/µL) and incubate reaction for 30 min at 37°C.
4. To precipitate the labeled oligonucleotide and remove unused $\gamma\,^{32}P$ ATP, add 25 µL 4 *M* ammonium acetate and 100 µL ethanol and leave at –20°C for at least 2 h. Alternatively, the oligonucleotide can be purified by using centrifugation through a gel-filtration column (*see* **Note 7**).

3.2. Amplification of Sample DNA

Six pmoles of upper and lower primer are used to amplify the DNA fragment of interest in a PCR in the presence of 0.2 m*M* dATP, dCTP, and dGTP, 0.02 m*M* dUTP and 0.19 m*M* dTTP, in a total volume of 20 µL (*see* **Note 1**). Normally, 100–200 ng genomic DNA sample serves as template for this reaction. It is recommended that a PCR of a normal or wild-type sample (and cleavage of same) is always included in each analysis for comparison purposes.

3.3. Glycosylase Mediated Cleavage of Amplified DNA

1. 10 µL of PCR is placed in 0.5-µL Eppendorf tube.
2. Add 1 µL Exonuclease I (0.5 U/µL) and incubate at 37°C for 30 min. Enzyme is diluted using dilution buffer.
3. The Exonuclease I is inactivated by incubating the reaction at 80°C for 15 min.
4. Put sample through clean-up column, e.g. S300HR (Pharmacia) or CentriSpin-10 (Princeton Seperations).
5. Take 10 µL of ExoI-treated DNA and add 1 µL uracil DNA-glycosylase (0.5 U/µL) and incubate at 37°C for 30 min.

6. Add 1 μL 0.6 *M* NaOH (final concentration of 0.05 *M* NaOH) and heat the reaction for 15 min at 95°C (*see* **Note 3**).
7. Add 1 μL 0.39 *M* Tris base (final concentration of 0.03 *M* Tris base) to neutralize the reaction.
8. Add 7 μL formamide loading dye to the sample and heat to 90°C for 5 min prior to loading on polyacrylamide gel.

3.4. PAGE

3.4.1. Manual PAGE and Autoradiography

1. Set up 6% denaturing polyacrylamide gel, containing 7 *M* urea and 1X TBE.
2. Wash out wells well with running buffer (1X TBE).
3. Load 5 μL of each sample (*see* **Note 8**).
4. Load labeled primer and uncleaved PCR product for use as size markers during fragment analysis.
5. Electrophorese at 60 W for several hours to get maximum separation of all fragments (*see* **Note 9** and **10**).
6. Expose the gel to X-ray film for several hours (*see* **Note 11**).
7. For additional detection of mutations, samples are run and analyzed on a 6% acrylamide (99:1) nondenaturing gel [similar to gel used for single-strand conformational polymorphism (SSCP) analysis *(6)*].

3.4.2. PAGE Using Alf Express (Pharmacia)

1. Make up a 6% denaturing PAG solution consisting of 42 g urea (7 *M*), 1.2X TBE (final concentration) and 15 mL of 40% acrylamide stock and bring volume to 100 mL with MilliQ grade H_2O.
2. Filter solution through at least a 0.45 μm filter.
3. Wash and set up large gel plates according to manufacturer's instructions.
4. Add 70 μL TEMED and 700 μL 10% APS to the gel solution and mix gently.
5. Pour gel according to manufacturer's instructions taking care to avoid the incorporation of air bubbles. Allow gel to polymerize fully for at least 2 h.
6. Set up gel in automated sequencer, wash out wells well with running buffer (0.6X TBE) and load 4 μL of sample per well.
7. Load labeled primer and uncleaved PCR product for use as size markers during fragment analysis.
8. Run gel at 55°C and at 35 W (constant) for several hours (*see* **Note 9**).
9. For additional detection of mutations, samples are run and analyzed on a 6% acrylamide (99:1) nondenaturing gel (similar to gel used for SSCP analysis) according to the manufacturer's instructions.

3.6.Fragment Analysis

1. Determine the position of the labeled primer on the gel.
2. All detectable cleaved fragments will be greater in size than this band.
3. Compare the pattern of cleavage fragments to that expected form the T content of the known sequence of the DNA being analyzed.

4. Each cleavage fragment should correspond to the sequence of the thymine residues in the known DNA sequence.
5. If the base change at the mutation site involves a gain of a T residue, then an additional band will be detected in the cleavage fragment pattern that does not correspond to the known DNA (T) sequence and may be present in the homozygous or heterozygous state. Whether the mutation exists in the homozygous or heterozygous state can only be determined using GMPD using comparative analysis between homozygous normal, heterozygous, and homozygous mutant samples.
6. If the base change at the mutation site involves a loss of a T residue, then one of the expected bands caused by the known T residue will not be detected.
7. If this loss of T mutation is present in a heterozygous state, then the intensity of the band due to the corresponding cleavage fragment will be reduced by 50%.
8. If a suitable size marker (labeled) is used during PAGE, then band size can be estimated by comparison to this marker either manually or automatically using fragment-analysis software.
9. The length or position of the gained or lost fragment is a direct indication of the location of the polymorphism or mutation.
10. GMPD using UDG and dUTP incorporation detects all deletion and insertion mutations. As 10 out of the 12 single-base substitution mutations involve the gain or loss of A/T or T/A basepairs (i.e., G>A, G>T, A>G, A>T, A>C, T>G, T>A, T>C, C>A, and C>T), 83% of single-base substitution mutations can be detected with UDG and dUTP. In reality, this figure is closer to 95% or greater of all point mutations because most disease-causing mutations are C to T (or G to A) transitions.
11. The two base changes not detected with UDG and dUTP incorporation, i.e., G to C and C to G occur at a low frequency relative to other base substitution mutations. These changes can be detected by using a modified precursor nucleotide which substitutes for dGTP or dCTP and a corresponding DNA glycosylase.
12. Alternatively, G to C and C to G changes are detected when the UDG generated cleavage products are analyzed under SSCP-like conditions. On such a gel, a banding pattern is observed that is somewhat similar to that observed when the cleavage products are analyzed on the denaturing sequencing type gel, however, the bands are often not as sharp or defined as those on the denaturing gel. Mutations are detected by the observation of additional bands in the mutant sample because of conformational changes in the those bands that alter their migration through the gel matrix.
13. GMPD offers the specificity of direct DNA sequencing, but is a much simpler process with almost all mutations detected using a single enzyme. The method does not rely on heteroduplex formation, but works equally well on homozygous and heterozygous samples. The annealing of primers is not dependent on stringent or differential hybridisation conditions. The amplification of samples using dUTP works well with several thermostable polymerases. The process can be performed directly on any DNA sample and the process is robust and requires relatively few manipulations.

4. Notes

1. The ratio of the dUTP to dTTP is such that approx one unit is incorporated per strand of amplified DNA. This ratio will vary depending on the base content of the DNA fragment being analyzed. In addition, the ratio required to analyze the two strands of any one fragment may be different. Therefore, it may be necessary to titrate the ratio of dUTP to dTTP required for analysis of a DNA strand by carrying out the aforementioned amplification and subsequent glycosylase-mediated cleavage using a range of ratios. Subsequently, a ratio is determined that is most informative for that particular DNA strand, i.e., the ratio that allows one to easily visualize a more or less uniform production of all expected GMPD cleavage fragments for that strand including the shorter fragments caused by cleavage of the amplified strands at the 5' end or close to the primer and longer fragments due to cleavage at the 3' end of the amplified strand. If all fragments cannot be efficiently and effectively produced with any one ratio, then it may be necessary to analyze the strand using more than one ratio.
2. Use of differentially labeled primers allows one to analyze multiple different DNA strands in each gel lane.
3. AP sites can be cleaved at the phosphate linkage on the 5' and/or 3' side of the ribose ring. Heating of the DNA at 95°C in the presence of alkali causes chemical cleavage at the 5' side of the AP site and leaves a phosphate group at the 3' terminus of the diagnostic fragment. AP endonuclease IV enzymatically cleaves at the 5' side of the AP site and leaves a 3'OH terminus on the diagnostic fragment. However, cleavage of AP sites by heating at 95°C at neutral pH results in chemical cleavage at the 3' side of the AP site, which leaves a deoxyribose phosphate moiety at the 3' terminus of the diagnostic fragment and a low level of cleavage at the 5' side leaving a 3' phosphate group on the diagnostic fragment. This results in the diagnostic fragment migrating as a doublet following denaturing PAGE. An AP lyase enzyme can also be used in place of the heat treatment because lyases also cleave at the 3' side of the AP site and result in a 3' deoxyribose phosphate terminus on the diagnostic fragment. The presence of this 3' deoxyribose phosphate group following 3' AP cleavage is a useful feature because the size of each fragment containing a 3' deoxyribose phosphate moiety is similar in size to that generated during a typical dideoxysequencing experiment, i.e., fragments generated from DNA containing random dUMP incorporation followed by UDG treatment and 3' AP cleavage are similar in size to the fragments generated during the dideoxythymine sequencing reaction (following eletrophoresis through a 6% denaturing gel). This allows for easy comparison of GMPD banding pattern with a sequencing pattern.
4. The oligonucleotide primers contain dG, dA, dT, and dC and consequently are resistant to cleavage by UDG.
5. When using the ALF Express for fragment separation and analysis (and/or use of Cy5 label on any fluorescence detection equipment), one must use a modified formamide loading dye containing dextran blue in place of bromophenol blue and xylene cyanol, because bromophenol blue and xylene cyanol fluoresce with 635 nm excitation and, hence, interfere with the detection of the Cy5 fluorescent label.

6. Do not autoclave TBE for use with the Alf Express.
7. Following 5' end labeling of the oligonucleotide, unincorporated γ^{32}PATP can be removed by ethanol precipitation or by using homemade gel filtration columns or any of the commercially available oligonucleotide purification kits (e.g., Centrispin-10 columns, Princeton Seperations, supplied by Genosys UK).
8. Normally, one may load 5 µL of sample per well. However, this may be altered depending on how well the oligonucleotide primer was labeled with ^{32}P and how long one is willing to leave the radioactive gel exposed to the X-ray film.
9. Gel running time depends on the expected size of fragments to be analyzed, percentage acrylamide in the gel, and thickness of the gel.
10. Multiple loadings depending on length of fragment being analyzed. The actual running time of the gel is determined by the size of the expected cleavage fragments. The progress of the fragments can be estimated by assuming bromophenol blue runs approximately as a 26 mer and xylene cyanol as a 106 mer in a 6% denaturing PAG.
11. A good starting point for X-ray film exposure is overnight exposure at –20°C.

References

1. Vaughan P. and McCarthy, T.V. (1999) Glycosylase mediated polymorphism detection (GMPD): a novel process for genetic analysis. *Genet. Anal. Biomol. Eng.* **14,** 169–175.
2. Lindahl, T. and Andersson, A. (1972) Rate of chain breakage at apurunic sites in double stranded deoxyribonucleic acid. *Biochemistry* **11,** 3618–3623.
3. Price, A. and Lindahl, T. (1991) Enzymatic release of 5'-terminal deoxyribose phosphate residues from damaged DNA in human cells. *Biochemistry* **30,** 8631–8637.
4. Lindahl, T., Ljungquist, S., Siegert, W., Nyberg, B., and Sperens, B. (1977) DNA N-glycosidases, properties of uracil-DNA glycosylase from *Escherichia coli*. *J. Biol. Chem.* **252,** 3286–3294 .
5. Duncan, B. (1981) in *The Enzymes* (Boyer, P., ed.), vol. 14, Academic, Orlando, FL, pp. 565–586.
6. Orita, M., Iwahana, H., Kanazawa, H., et al. (1989) Detection of polymorphisms of human DNA by gel electrophoresis as single strand conformation polymorphisms. *Proc. Natl. Acad. Sci. USA* **86,** 2766–2770.

16

The Use of Resolvases T4 Endonuclease VII and T7 Endonuclease I in Mutation Detection

Jeffrey J. Babon, Matthew McKenzie, and Richard G. H. Cotton

1. Introduction

The use of resolvase enzymes to detect mutations *(1)* was developed in response to the demand for a method that could screen kilobase lengths of DNA for single nucleotide changes and small insertions and deletions. The method is a more simple, nontoxic alternative to the chemical cleavage of the mismatch (CCM) method *(2)* and as such, the techniques proceed along similar lines; heteroduplex formation then mismatch cleavage then electrophoresis to visualize the reaction products. Both techniques use as substrate, heteroduplexes generated by the melting and reannealing of query and control DNA, usually polymerase chain reaction (PCR) products, in the same tube. If the sequence of the two original DNA species differ at any nucleotide, then heteroduplex species will be generated with a base-pair mismatch at that position (*see* **Fig. 1**). These mismatches can be bound and the DNA cleaved by at least two resolvase enzymes, T4 endonuclease VII *(1)* and T7 endonuclease I *(3)*, both bacteriophage enzymes with similar in vivo functions. The one-step binding and cleavage reaction replace the two-step CCM procedure that uses different chemicals in each stage that are not active in the same buffer and thus require a clean-up step in between. Another advantage of using resolvases rather than chemicals for this procedure is that the one enzyme can recognize all eight types of mismatches (a/a, t/t, c/c, g/g, a/g, t/g, a/c, t/c) *(4)*, as well as small loops generated when the query DNA used to form the heteroduplex substrate contains a small insertion or deletion *(5)*. The bacteriophage resolvases, T4 endonuclease VII and T7 endonuclease I, have as their primary in vivo purpose, the resolution of Holliday junctions *(6,7)*, four-stranded DNA junctions that are formed during the phage-replication process. These enzymes

From: *Methods in Molecular Biology, vol. 152: DNA Repair Protocols: Prokaryotic Systems*
Edited by: P. Vaughan © Humana Press Inc., Totowa, NJ

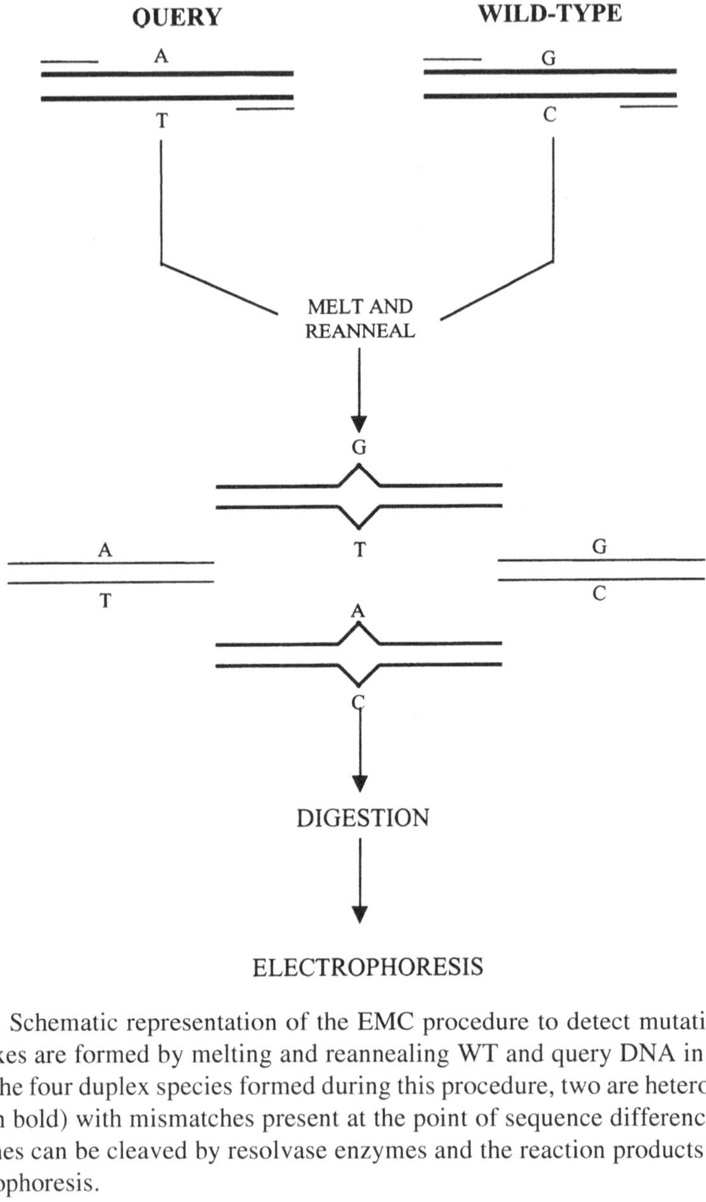

Fig. 1. Schematic representation of the EMC procedure to detect mutations. Heteroduplexes are formed by melting and reannealing WT and query DNA in the same tube. Of the four duplex species formed during this procedure, two are heteroduplexes (shown in bold) with mismatches present at the point of sequence differences. These mismatches can be cleaved by resolvase enzymes and the reaction products analyzed by electrophoresis.

were later found to cleave a range of three-dimensional DNA structures. T4 endonuclease VII, in particular, has had its substrates well characterized and the structures recognized as substrates include DNA mismatches, loops, 4- and 3- stranded junctions, gaps, overhangs *(8)*, as well as sequence-induced bends, such as those resulting from phased A-tracts *(9)*. T4 endonuclease VII or T7 endonuclease I bind to the DNA and then cleave within six nucleotides 3' or 5', respectively ***(10,11)***. A counternick will then be placed on the other side of the mismatch on the opposite strand again within six nucleotides. The two nicks are not necessarily symmetrical about the mismatch and often have very different reaction rates ***(4)***. Occasionally, only one nick will ever be detected, therefore, the strands are separated and the reaction products are viewed after electrophoresis through a denaturing polyacrylamide gel so that single-stranded cleavages will be detected, as well as double-stranded breaks. The structures recognized as substrates for these enzymes, as well as the mechanism of cleavage, place these enzymes in a class distinct from mismatch repair enzymes that will often cleave only some types of mismatches or in only certain sequence contexts or else cleave at a large distance from the mismatch [for review, *see **(12)***]. The nature of resolvase cleavage means that all mismatch types are cleaved in almost all sequence contexts and the DNA is cleaved close to the mismatch, allowing the mutation to be positioned within 6 bp. The general nature of substrate recognition, however, is responsible for a certain "background" cleavage of homoduplex regions of DNA that in certain cases can be more intense than the mismatch cleavage product itself. This is possibly caused by sequence induced secondary structure in the DNA, but the problem is resolved by cleaving a homoduplex control DNA, any identical peaks in the control and sample lanes are designated as background and the mutation band is usually easily distinguished (*see* **Fig. 2**). When this precaution is taken, nearly 100% of mutations can be detected ***(13)***.

2. Materials

T4 endonuclease VII and all necessary buffers are now commercially available from Amersham Pharmacia as part of the Passport™ EMD kit, the protocol used with this kit differs from that shown here, so follow manufacturer's instructions (*see* **Note 1**).

1. TE buffer: 10 m*M* Tris-HCl, pH 8.0, 1 m*M* ethylenediaminetetraacetic acid (EDTA).
2. 2X annealing buffer: 1.2 *M* NaCl, 14 m*M* $MgCl_2$, 12 m*M* Tris-HCl, pH 8.0
3. T4 endonuclease VII or T7 endonuclease I, dilute fresh before each experiment to working concentration, units will differ between suppliers and preparations therefore refer to manufacturer's instructions or quantitate activity yourself on a reference sample.
4. 10 reaction buffer (supplied) or: 500 m*M* Tris-HCl, pH 8.0, 100 m*M* $MgCl_2$, 1 mg/mL bovine serum albumin (BSA), 50 m*M* dithiothreitol (DTT).
5. Dilution buffer: 10 m*M* Tris-HCl, pH 8.0, 0.1 m*M* glutathione, 50% glycerol, 0.1 mg/mL BSA.
6. Loading buffer: 95% formamide (toxic), 10 mg/mL blue dextran.
7. 1X kinase buffer: 50 m*M* Tris-HCl, pH 7.6, 10 m*M* $MgCl_2$, 5 m*M* DTT, 0.1 m*M* spermidine, and 0.1 m*M* EDTA.

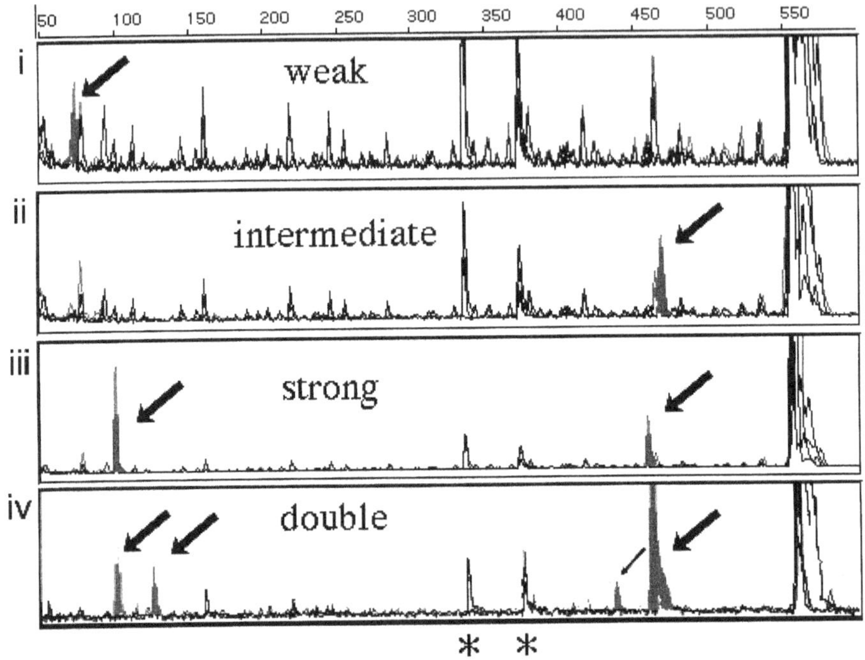

Fig. 2. Results of fluorescent EMC. Examples of weak **(i)**, intermediate **(ii)**, and strong **(iii)**, cleavage of different mismatches in the same sequence context by T4 endonuclease VII as well as two different mutations detected in the one fragment **(iv)**. T4 endonuclease VII was incubated with the different heteroduplexes and a homoduplex control in the same sequence context under standard conditions. The results of the cleavage are shown after electrophoresis and analysis on an ABI-377 sequencer. Each panel shows the result of a heteroduplex cleavage reaction (gray) overlaid on to the homoduplex control reaction (black) Specific cleavages are marked with arrows and are colored gray, the two major background bands in this fragment, present in all heteroduplexes as well as the homoduplex control, are labeled with asterisks. Approximate sizes in bp are shown above. Note that in the case of weakly recognized mutations **(i)** the specific band can be less intense than the major background band whereas for strongly recognized mutations **(iii)** the mutation bands dwarf the background bands. Also note that not all mutations, e.g., **(i)** and **(ii)**, show two cleavage products for the two-strand cleavages.

3. Methods

Note: This procedure can be performed fluorescently ***(14)*** or radioactively ***(1,15)***. Fluorescent EMC is the method of choice (*see* **Note 2**) if the user has access to an ABI-377 sequencer or analogous machine therefore the fluorescent protocol is given as follows. For radioactive EMC, *see* **Note 2**.

3.1. DNA Preparation

Amplify target sequence from wild-type/control (WT) and query DNA using standard PCR protocols and fluorescent primers, alternatively restriction enzyme digested DNA can be used and the ends labeled according to standard protocols with a fluorescent nucleotide (*see* **Note 3**). The fragment size should be kept at or below 1 kb (*see* **Note 4**). Twenty-five nanograms of each species per resolvase reaction is required with 2X 25 ng for the WT homoduplex control.

Purify the PCR products using standard methods such as spin-column purification (Qia-quick spin columns, Qiagen, Germany) or gel electrophoresis, followed by band excision and DNA extraction (e.g., Bresaclean, Bresagen Australia), following manufacturer's instructions (*see* **Note 5** for possible omission of this step), resuspend the DNA in TE buffer.

3.2. Heteroduplex Formation

1. Place into a tube query DNA and WT DNA in a 1:1 ratio if both species are labeled or in a 10:1 ratio if only WT is labeled.
   There should be 25 ng WT DNA per reaction and either 25 or 250 ng query DNA depending on whether it is labeled or unlabeled respectively (*see* **Note 6**), the heteroduplex formation reaction can be scaled up if multiple reactions are to be performed on the one substrate but the volume should not exceed 200 μL (*see* **Note 7**).
2. Add 2X annealing buffer or 10X PCR buffer to 1X.
3. Boil for 5 min.
4. Place the tube in 65°C water bath for at least 30 min (*see* **Note 7**).
5. Place the tube at room temperature for 30 min.
6. Place tube on ice.

3.3. Resolvase Cleavage Reaction

The heteroduplexes are now ready to be cleaved. Each reaction should contain 50 ng of heteroduplex DNA and should be performed in a 10–50 μL reaction volume. The heteroduplex DNA added must be of a volume low enough that the sodium chloride or other monovalent salt concentration is diluted to 50 m*M* or less.

1. Add the appropriate amount of water and 10X reaction buffer to the heteroduplex DNA and finally, add the appropriate amount of freshly diluted enzyme (*see* **Notes 8–11**) and incubate at 37°C for 30 min.
2. Stop the reaction by adding 2 vol of ice-cold ethanol and 1/10 vol of 3 *M* sodium acetate, pH 5.6 (*see* **Note 12** for alternative).
3. Incubate on ice for at least 30 min.
4. Centrifuge at 13,000*g* for 10 min in a microcentrifuge, decant the supernatant.
5. Wash the pellet once with 500 μL 70% ethanol and repeat **step 4**.
6. Dry the pellet and resuspend in loading buffer. The samples can then be analyzed by electrophoresis on an ABI 377 sequencer or similar instrument.

7. For analysis on the ABI 377, electrophoresis is performed on a 4.25% (19:1) acrylamide:*bis*-acrylamide, 6 *M* urea gel, run in TBE buffer at 3000 V. A 500 bp fragment will take 3 h to run in this fashion. The matrix used depends on the primer label, matrix C will detect 6-FAM, HEX, and TET-labeled DNA and a TAMRA-labeled size marker. Matrix D will detect 6-FAM, ROX, NED, and TAMRA-labeled DNA.

3.4. Results Analysis

Print out the results of each query DNA reaction overlaid with the homoduplex control reaction. Cleavage peaks present in the query trace and not the control represent differences in the sequence near that position. Any peaks present in both that are substantially more intense in the query trace may also represent a mutation (*see* **Note 13–17**).

4. Notes

1. *The Amersham Pharmacia PassportTM EMD kit.* The background cleavage activity of the enzyme, although annoying, does not usually inhibit confident mutation detection when using resolvase cleavage. A bigger drawback to the technique is the number of manipulations required from start to finish. Amersham Pharmacia have recently released a kit form of the EMC technique, called Passport. This kit has been developed to increase the speed and ease of the original EMC protocol. The kit is provided with optimized buffers and reaction conditions for the heteroduplex formation, resolvase cleavage reaction, and gel-loading steps. These conditions allow the elimination of the final ethanol precipitation step, and the post-PCR purification step while still giving a sufficient signal-to-noise ratio to detect mutations. When using this kit, refer to the manufacturer's instructions.
2. *Radioactive EMC.* Label the WT PCR product in bulk using γ-^{32}P-ATP (setting aside 250 ng unlabeled for the homoduplex control) as follows.

Incubate the DNA for 60 min at 37°C with 1 U PNK (New England Biolabs) and 1 µL γ-^{32}P-ATP (10 mCi/mL, 3000 Ci/mmol) in 1 × PNK buffer (supplied). The amount of radiolabel added should be adjusted to result in around 100 cpm per ng DNA.

Purify the labeled DNA from the remaining γ ^{32}PATP by ethanol precipitation (2.5 vol ice-cold ethanol, 0.1 vole 3 *M* sodium acetate, pH 5.2. Leave on ice for 30 min then pellet DNA by centrifugation at 13,000*g* for 10 min. Perform two ice-cold 70% ethanol washes with a 2 min centrifugation step to pellet the DNA in each case. Resuspend the DNA at 50 ng/µL in TE buffer). The ethanol precipitation step can be replaced by using a spin column such as the Qia-quick nucleotide removal kit (Qiagen, Germany) according to manufacturer's instructions. Resuspend the DNA in TE buffer. Form heteroduplexes and perform digestion as usual, stop by adding an equal volume of loading buffer containing 25 m*M* EDTA and electrophorese on the appropriate percent polyacrylamide gel, visualize by autoradiography.

Electrophoresis of samples under 1 kb is performed on an 8 *M* urea, 6% (19:1) acrylamide/*bis*-acrylamide gel in 1X TBE buffer at 45 W. We routinely use gels that are 420 × 210 × 0.4 mm in size. Often multiple loadings will be necessary to ensure the full fragment is analyzed. Otherwise, larger fragments remain unresolved near the top of the gel. Typically, 90 min will cover the range up to approx 300 bp with longer times required for 300–1000 bp fragments. Bromophenol blue and xylene cyanol dyes should be added to the loading buffer (at 0.1%) to keep track of the position of varioussized fragments migrating thrrough the gel. In a 6% denaturing gel, bromophenol blue migrates approx as a 26-bp fragment whereas xylene cyanol migrates as a 106-bp fragment. After electrophoresis, the gels are fixed in 10% acetic acid, dried for 1 h and placed in a cassette with Kodak X-OMAT AR film overnight.

Radioactive EMC is inferior to fluorescent EMC for a number of reasons, relating more to the whole process than to radioactivity vs fluorescence per se. The near linear relationship between size (bp) and time detected (i.e., time taken for a fragment to migrate past the laser) on an ABI-377 sequencer means that only one run is necessary per fragment. As shown in **Fig. 3**, if the run conditions in radioactive EMC are optimal for detecting smaller cleavage fragments, then the larger fragments are often lost in the high concentration of bands at the top of the gel. In addition to this, using a different fluorescent label on each end allows each mutation cleavage band to be attributed to a position on the correct DNA strand. Radioactive EMC has the problem of the background cleavage on the two strands being additive, when each strand is a different color this problem is alleviated and the signal to noise ratio increases. Finally there is no extra labeling step required with fluorescent EMC although fluorescent primers do need to be purchased. Of course not all laboratories have access to an ABI-377 or similar instrument and radioactive EMC despite its limitations is more than adequate to detect mutations.

3. *End vs uniformly labeled heteroduplexes.* Ensure that the fragment is end rather than uniformly labeled. Nick-translation labeling or incorporation of label during PCR is not recommended. End labeling ensures that each nick in the DNA, either specific or background, has only one signal.
4. *Fragment length.* Lengths longer than 1 kb may make it too difficult to distinguish the specific signal from the background signals and also tend to make heteroduplex formation more difficult.
5. *DNA purification.* Purification post-PCR is omitted when using the Amersham Pharmacia Passport kit. However, purification leads to consistently better results in our hands when using raw enzyme and materials. When using unpurified PCR

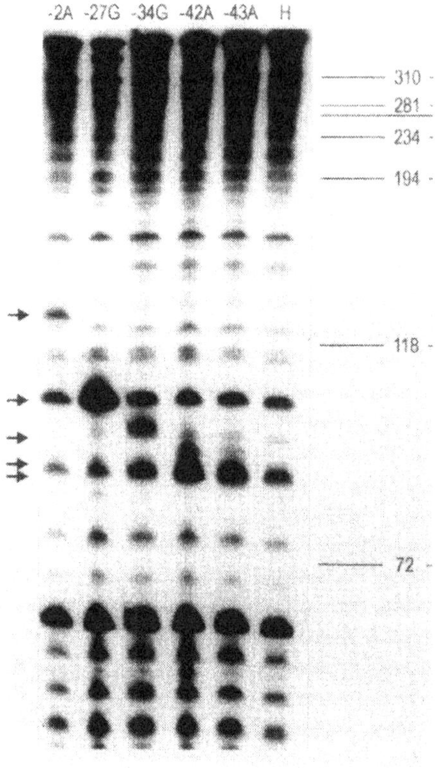

Fig. 3. Results of radioactive EMC. Five different mutants and a homoduplex control in the same 546 bp sequence context were analyzed for mutations using T4 endonuclease VII cleavage. The heteroduplexes were formed by annealing radiolabeled WT DNA and cold mutant DNA and the reaction products visualized after electrophoresis by autoradiography. Sizes in bp are shown to the right. The specific cleavages are marked with arrows and are, from right to left, 118 bp, 99 bp, 91 bp, 82 bp, and 81 bp, whereas the cleavage bands present in all lanes are examples of background cleavage. The high level of background cleavage with these fragments illustrates the need for a homoduplex control in each experiment. Also note that the large fragment for each mutation is not easily visible because of the high concentration of bands at the top of the gel.

products ensure that a zero enzyme control for each heteroduplex is performed so that PCR contaminants are not mistaken as cleavage products

6. *Wild-type to mutant ratio.* Using excess unlabeled query DNA yields the lowest background cleavage, as the labeled WT DNA strands are more likely to form heteroduplex species with the excess query strands during the melting/re-annealing procedure, rather than reannealing to form WT homoduplexes again. This is usually convenient anyway because often the WT DNA can be prepared in bulk and

25-ng aliquots annealed to entire query PCR products to form heteroduplexes. However, this means that only two of four heteroduplex strands are being analyzed and occasionally this can lead to a mutation not being detected. Theoretically, near 50% of the duplexes formed by melting and reannealing WT and query DNA in a ratio of 1:1 will be heteroduplexes and this is sufficient for the mismatch signal to be distinguishable from the background. When both species are labeled then all four strands are examined. A 1:1 WT:query DNA ratio is advised in this case.

7. *Heteroduplex formation.* There is a lot of freedom in the conditions allowing good heteroduplex formation, however, large volumes introduce the problem of inefficient heating of the samples during the denaturing step and should be avoided. The matter of optimum DNA concentration is unresolved, too low and too high a concentration have been reported to lead to various problems ***(16)***, however, we consistently use DNA concentrations ranging from 1–40 ng/µL. The matter of the length of time at 65°C is another unresolved issue in heteroduplex formation. 65°C is on the melting curve of duplex DNA and will ensure that the temperature is high enough so that any hairpin loops or other single-stranded structures formed while cooling down from 95°C will be able to unwind so that duplex formation can occur. Some protocols with troublesome DNA sequences suggest an overnight incubation ***(17)*** at 65°C, but this is seldom necessary. To examine the heteroduplex formation reaction, run an aliquot on an agarose gel and ensure that the major band is of the correct size and that multiple intense bands or smeary bands are not present ***(16)***. Sometimes multistranded DNA complexes or excess single-stranded DNA may be formed and this can lead to an increased background cleavage level ***(16)***.
8. *Enzyme stability.* The enzyme is quite stable in storage buffer at –20°C, however, we recommend not storing the enzyme at working concentration for more than a few weeks as there is a suggestion that in more dilute solutions the enzyme can lose activity over time (J. Babon, unpublished data). It should not be repeatedly freeze-thawed at –70°C. The enzyme is inhibited by monovalent cations such as K+, Na+ above 50 m*M*, many divalent cations above 10 m*M* and by Zn2+ above 1 m*M*, its activity is minimal above 45°C and is inactivated by incubations at any temperature above 30°C when substrate (DNA) is not present. All these parameters should therefore be avoided.
9. *Total digestion.* In our experience, the total duplex digestion should be adjusted so that 50–90% of the substrate band remains. The total cleavage using the same amount of substrate and enzyme will differ to some degree depending on mutation(s) present and the sequence of the fragment. The enzyme is unlike a restriction enzyme and is capable of entirely digesting a fragment of DNA eventually whether mismatches are present or not. However, this will only happen if a large excess of enzyme is used and it is wise to quantitate the effects of the enzyme on the particular fragment of interest before a major study is undertaken. As the incubation continues, the shorter fragments become more intense, whereas the larger fragments are further digested and diminish. In some cases, multiple time-points (5 and 30 min, e.g.) may be beneficial. Overdigestion can cause the

background cleavage products to accumulate to a level sufficient to obscure a mutation, whereas underdigestion can result in a weak mutation being undetected.

10. *Cleavage characteristics.* The kinetics of cleavage of each mutation differ depending on mismatch type and sequence context. The enzyme's activity on different mismatches has been, in the past, classified into strong, intermediate, and weak ***(13)***. However, in reality, the activity can cover the full range between unreactive and very rare ***(3)*** to acting like a restriction enzyme for mismatches. In many cases the total cleavage of a heteroduplex containing a well-recognized mismatch will be much greater than a homoduplex control, reflecting the affinity of the enzyme for that mismatch. However, many poorly recognized mismatches will show an insignificant difference in the level of total digestion compared to the homoduplex control. The other characteristics of cleavage by this enzyme is the number of nicks placed in each strand. We have observed as many as three nicks placed within the six nucleotide limit of cleavage at a single mismatch, therefore, observing such a phenomena does not necessarily indicate the presence of three mutations.
11. *Endonuclease VII and endonuclease I.* At present, only one study has been published using both T4 endonuclease VII and T7 endonuclease I. This study and our experience suggests that both enzymes are equally effective in detecting mutations. They both detect close to 100% of mutations ***(1,3,13)***. Both will recognize all types of mismatches and small heteroduplex loops and with similar signal-to-noise ratios, although the background cleavages are in completely different positions (*see* **Fig. 4**).
12. *Stopping the reaction.* Instead of ethanol precipitation and running the whole sample on the gel, an aliquot of the digested sample can be run directly on to the gel, which speeds up the process. The 0.2-mm ABI-377 sequencer gels cannot tolerate even a moderate salt concentration in the sample without problems. However, if using the thicker manual sequencing polyacrylamide gels, this approach can be used and, in this case, the reaction can be stopped by adding an equal volume of loading buffer and EDTA to 25 m*M*.
13. *Identifying mutations.* Mutations are identified by peaks either not present at all in a homoduplex control or else present, but as a much lower contribution to the total cleavage. A mutation peak at the same position as a major background peak represents the most difficult mutation to detect. In this case, one is looking for an increase in the intensity of that peak compared to the other background peaks present in the query trace relative to the same comparison in the control trace. If all the peaks present in the query trace compared to the control are elevated, this will usually mean that there has been greater total digestion in the query reaction compared with the control and is not representative of mutations.
14. *Single- vs double-stranded cleavage.* Most mutations will be represented by two peaks, one for each DNA strand. However, many mutations will only be represented by one peak as the enzyme will not always cleave both strands at a mismatch to an extent where they are both detectable. The enzyme's activity upon the four mismatched bases in the two heteroduplex species present in each reaction appears to be independent such that the nicking activity upon the four strands can differ greatly in intensity. In some very rare cases that we have dealt with,

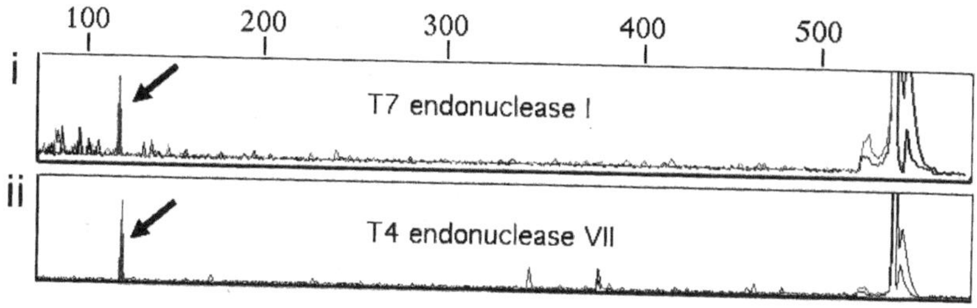

Fig. 4. Detecting the same mutation with T4 endonuclease VII and T7 endonuclease I. A 546 bp DNA fragment with a mismatch located 118 bp from the 5' end of the sense strand were digested with T7 endonuclease I (**i**) and T4 endonuclease VII (**ii**). Approximate sizes in bp are shown above. In the case of this mutation, only one of the two cleavage fragments is produced by incubation with either enzyme. T7 endonuclease I cleaves on the opposite side of the mismatch to T4 endonuclease VII and hence the cleavage fragment is approx 6 bp smaller. Background bands result after incubation with either enzyme but are located in completely different positions.

when only two of the four strands were labeled, no cleavage products could be seen, but when the opposite two strands were labeled a cleavage product was visible. This, of course, is an argument for labeling all four strands.

15. *Multiple mutations.* Multiple mutations in the one fragment can be detected using this technique as long as digestion is not overextensive. Because the DNA is end labeled, overdigestion may result in the larger cleavage fragments disappearing as they are recleaved at the position of a second or subsequent mutation. Usually, this is not a problem if the total digestion recommendations are followed.
16. *Screening criteria.* To be pronounced free of mutations within the limitations of the method, a test heteroduplex must show no peaks that are not present at a similar level in the homoduplex control. In addition, it must be ensured that the total digestion of both homoduplex control and sample are of similar magnitude (note that often, but not always, the existence of a mismatch will cause the fragment to be cleaved significantly more quickly, *see* **Note 10**). All four strands must be examined for cleavage and if electrophoresis is performed on a regular sequencing-gel apparatus, then a long and short run should be performed to ensure that small and large fragments are visualized.
17. *Increasing the signal-to-noise ratio.* In rare cases, the background cleavage of particular DNA fragments may inhibit confident screening by resolvase cleavage. Often this is caused by a high level of homoduplex DNA still present after the heteroduplex formation reaction. An easy method of purifying heteroduplexes has been published previously *(18)* that requires the use of biotinylated primers to PCR amplify either the mutant or the WT DNA. The resulting heteroduplexes will have a biotin and a fluorescent moiety on opposite strands and can be thus purified from the homoduplex DNA by using streptavidin-labeled beads and the

reactions performed on the solid phase. This has been shown to greatly increase the signal-to-noise ratio of the technique. However, it does require an extra pair of biotinylated primers to be purchased and only two of the four heteroduplex strands will be fluorescently labeled. Therefore, the use of this adaptation is only recommended if the background cleavage of the DNA fragment of interest is very high such that mutation peaks cannot be identified with confidence.

References

1. Youil, R., Kemper B. W., and Cotton R. G. H. (1993) Screening for mutations by enzyme mismatch cleavage using T4 endonuclease VII. *Am. J. Hum. Genet.* **53,** abst. 1257.
2. Cotton R. G. H., Rodrigues, N. R., and Campbell, R. D. (1998) Reactivity of cytosine and thymine in single-base-pair mismatches with hydroxylamine and osmium tetroxide and its application to the study of mutations. *Proc. Natl. Acad. Sci.* **85,** 4397–4401.
3. Mashal, R. D., Koontz, J., and Sklar, J. (1996) Detection of mutations by cleavage of DNA heteroduplexes with bacteriophage resolvases. *Nat. Genet.* **23,** 177–183.
4. Solaro, P.C., Birkenkamp, K., Pfeiffer, P., and Kemper, B. (1993) Endonuclease VII of phage T4 triggers mismatch correction in vitro. *J. Mol. Biol.* **230,** 868–877.
5. Kleff, S. and Kemper, B. (1988) Initiation of heteroduplex-loop repair by T4-encoded endonuclease VII in vitro. *EMBO J.* **7,** 1527–1535.
6. Kemper, B. and Garabett, M. (1981) Studies on T4-Head maturation. *Eur. J. Biochem.* **115,** 123–131.
7. Dickie, P., McFadden, G., and Morgan, A. R. (1987) The site-specific cleavage of synthetic Holliday junction analogues and related branched DNA structures by bacteriophage T7 endonuclease 1. *J. Biol. Chem.* **262,** 14,826–14,836.
8. Kemper, B., Pottmeyer, S., Solaro, P., and Kosak H. (1990) Resolution of DNA-secondary structures by endonuclease VII (endo VII) from phage T4, in *Human Genome Initiative and DNA Recombination,* Vol. 1, Structure and Methods, Adenine Press, New York.
9. Bhattacharyya A., Murchie A. I. H., von Kitzing E., et al. (1991) Model for the interaction of DNA junctions and resolving enzymes. *J. Mol. Biol.* **221,** 1191–1207.
10. Parsons, C. A., Kemper, B., and West, S. C. (1990) Interaction of a four-way junction in DNA with T4 endonuclease VII. *J. Biol. Chem.* **265,** 9285–9289.
11. de Massey, B., Studier, F. W., Dorgai, I., Appelbaum, E., and Weisberg, R. A. (1985) Resolution of Holliday Structures by Endonuclease VII as observed in interactions with cruciform DNA. *Cold Spring Harbor Symp. Quant. Biol.* **49,** 715–726.
12. Modrich, P. (1991) Mechanisms and biological effects of mismatch repair. *Ann. Rev. Genet.* **25,** 229–253.
13. Youil, R., Kemper, B., and Cotton, R. G. H. (1996) Detection of 81 of 81 known mouse b-globin promoter mutations with T4 endonuclease VII-the EMC method. *Genomics* **32,** 431–435.
14. Del Tito, B. J., Jr., Poff, H. E. 3rd, Novotny, M. A., et al. (1998) Automated fluorescent analysis procedure for enzymatic mutation detection. *Clin. Chem.* **44,** 731–739.

15. Youil, R. and Cotton, R. G. H. (1996) EMC-enzyme mismatch cleavage, in *Laboratory Protocols for Mutation Detection,* Oxford University Press, Oxford, UK, pp. 65–68.
16. Yakubovskaya, M. G., Humphrey, K. E., Babon, J. J., et al. (2000) Phenomenon of diverse DNA structures appearing in concentrated DNA solutions after purification or heteroduplex formation procedure. *Nucleic Acids Res., Science*, submitted.
17. Biasotto, M., Meo, T., Tosi, M., and Verpy, E. (1996) FAMA fluorescence-assisted mismatch analysis by chemical cleavage, in *Laboratory Protocols for Mutation Detection,* Oxford University Press, Oxford, UK, pp. 54–60.
18. Babon, J. J., Youil, R., and Cotton, R. G. H. (1995) Improved strategy for mutation detection-A modification to the enzyme mismatch cleavage method. *Nucleic Acids Res.* **23,** 5082–5084.

Index

FSC®
www.fsc.org
MIX
Papier | Fördert
gute Waldnutzung
FSC® C083411

Kolibri 360 GmbH
Ferdinand-Jühlke-Straße 7
99095 Erfurt, Deutschland
produktsicherheit@kolibri360.de